HIRED MINDS

HIRED MINDS

A Career Guide for Engineering Students and Graduates

Bryan Gardner

AIAA

American Institute of Aeronautics and Astronautics, Inc., Reston, Virginia

1 2 3 4 5

Library of Congress Cataloging-in-Publication Data

Gardner, Byan.
Hired minds: a career guide for engineering students and graduates / Bryan Gardner.
p.cm.
ISBN 978-1-56347-876-5 (alk. paper)
1. Aeronautics--Vocational guidance. I. Title.

TL561.G37 2007
620.0023--dc22

2007018163

TABLE OF CONTENTS

Acknowledgments

First, a shout goes out to my good engineering friends for their superb editorial skills: Katie Timothy, Brittany Reed, and Amber Hardy. It's great to have friends who will compliment you, but it's even better to have friends who will tear into your work with ruthless criticism, taking no concern for your emotional well being. Katie's point blank suggestions really helped to pinpoint what to include and what to cut. Brittany's natural sense of organization significantly improved the flow of several sections. Amber fixed more grammatical rules than I even knew existed. They may not have fixed all my stylistic quirks, but they certainly smoothed some very rough edges.

Loretta Hidalgo triggered me into action at a few critical points, and she followed up for results. Her energy is inspiring, her support is genuine, and she's a shining example of living life to its full potential.

I want to thank the professors at Brigham Young University who have been very encouraging and helpful with this process, especially Dr. Robert Todd and Dr. Lloyd Hawkins. A little seasoned advice and guidance goes a long way. In addition, Janie Hansen-Nelson and all the secretaries in the Mechanical Engineering department were indispensable in arranging the resume workshops from which this book started.

Thanks to all the engineers and managers at Northrop Grumman, Azusa, who have been great to work with and made my first few years as an engineer an excellent experience. Laura Fishman has been the best manager I could have asked for. She has trusted me with projects that could have been over my head, and given me the support to succeed. The flexibility she allowed has been very helpful in doing all the things outside work that are equally important to enjoying life. I can't overlook the influence of Craig Snow. His objective and candid manner reinforces strong performance and is instructive in missteps. He has made several opportunities available, without which this book would have never been started. Paul Russell proactively took the time to review the manuscript when I thought it was ready for publication. His perspective as an executive made a significant difference in the final product and triggered the final round of reviews that really completed the book.

Greg Larson was an indispensable connection to the people at AIAA. From workshops with Los Angeles area universities to providing contacts, he was always willing to help out in any way possible.

My brothers and parents have been far more influential than they may think. This book is about living life to the fullest, and they're my strongest examples of doing just that.

Whether laughing about a screwball medical patient, ripping around in an airplane, teaching in Mexico, or pulling turns out waterskiing, it's encouraging to see hard work and fun go together. My mom's improvisational throw-it-together-and-make-it-work style is the example I followed in starting and finishing this book, while my dad's structured and methodical work habits carried over to keep everything organized throughout the process. I couldn't ask for a better family.

And finally, thanks to all my friends and the engineers whose experiences make up the content of this book. There are several whose experiences were influential in developing the principles in this book, but aren't included for the sake of brevity. Your names and experiences may not be in here directly, but they helped to shape the content and flow. Thanks for the conversations, and most of all, thanks for making it fun.

CHAPTER 1

Introduction

June 21, 2004

On June 21, 2004, as Burt Rutan and Paul Allen's SpaceShipOne rocketed from the Mojave Desert to the edge of space, 63 miles in the sky, with pilot Michael Melvill earning the title of the first private astronaut to enter space, a whole tarmac full of engineers, scientists, pilots, and space enthusiasts cheered. Thousands on the ground watched the white streak of rocket fuel cut a line through the rising sun, straight to the top of the cloudless blue sky. The view from the spacecraft held the blackness of space, the curvature of the earth, and the whole southern half of California. Have you ever seen the earth from that altitude? Few have, but as the engineers of this world keep working toward their goals, many more will own that experience.

Engineering is the process of harnessing the elements, organizing them to our design, and orchestrating them to create the form and motion envisioned in our minds. Engineering is not dreaming. As engineers, we forge reality according to our design. However, along the process we frequently find it is our design that is forged by reality.

The Universe operates on the laws of nature, and the laws of nature do not bend. If you set the force of your will against the force of nature, you will lose every time. However, if you apply your mind to understand the forces of nature, you can leverage nature against itself, and accomplish your goals.

Burt Rutan has an impeccable history of engineering cutting edge aircraft. Among the more prominent of his designs is the Voyager aircraft, the first to fly around the world, non-stop, non-refueled, and more than doubling the existing range record on its nine-day flight. Following on that record, he designed the Global Flyer, repeating a non-refueled flight around the world, but this time with a solo pilot accomplishing the trip in 67 hours. He has also been highly successful designing composite aircraft for commercial markets.

Burt has applied his mind and learned to work with the laws of nature, yet even he realizes that not everything is foreseen and accounted for in design. Shortly after the historic first flight of SpaceShipOne, he told a group of us about some of the challenges they came across in engineering this project. With the spacecraft still on the runway, he pointed out two small fins on the body, about 4" by 8", and started explaining, "Because I failed to design the aircraft properly, we found during testing that we had to add these fins to get the stability we needed." Even to add those small fins, they needed to understand the airflow around the body, and design the right fin, in the right location, at the right angle. Only then could they correct their design.

"Question. Never Defend." Burt Rutan and his engineering team live by this design policy. They don't rely on luck to make their designs work. They set high goals for their objectives, and then put in the work necessary to reach their goals.

Burt Rutan and his team of engineers weren't the only people involved in this revolutionary space flight. Like all achievements of historical significance, there are several individuals who are indispensable to success.

Paul Allen, cofounder of Microsoft, is one of the key movers responsible for SpaceShipOne. Paul has been involved with space for many years, and founded Space Adventures, the company that made arrangements for the civilian space explorers the Russians have taken to the International Space Station.

Paul's interest in developing a suborbital spacecraft was inspired by the X Prize, a $10 million award for the first private suborbital manned space flight to be accomplished and repeated within two weeks. The X Prize was created by space entrepreneur Peter Diamandis, and was only the beginning of several manned space flight competitions to follow.

When Paul heard about the X Prize, he began investigating the developmental aircraft and rocket industry. He was intent on finding the best engineers who could successfully design, build, test, and fly the mission. He initiated contact with Burt Rutan, committed himself to funding the project, and the rest is history.

Mike Melvill was 61 years old piloting this flight, adding to his collection of nine world and national speed and altitude records. He has been one of the most respected commercial test pilots for decades, logging time on 128 fixed wing and 12 rotary wing aircraft. He has and even built his own airplane, and flown it around the world with fellow test pilot Dick Rutan, Burt Rutan's brother.

These aren't the heroes of yesterday. These people are the heroes of today. I had to laugh as I was standing in the crowd and watched moon walker Buzz Aldrin (Apollo 11), white haired but still alive and strong, work his way to the front of the crowd so that he could shake hands with Mike Melvill. You know something significant has happened when a man who is continually being asked for a handshake and a picture is now moving forward, asking for a handshake and a picture with a new hero.

The accomplishments of Charles Lindberg, John Glen, and Neil Armstrong are now history, but the way that they lived is still alive and well. Pilots like Chuck Yeager are followed by Mike Melvill. Aircraft designers like Jack Northrop are followed by Burt Rutan. Sponsorship has expanded to include independent entrepreneurs and financiers like Peter Diamandis and Paul Allen. The heroes of yesterday have made way for the heroes of today, and those driving the accomplishments of today will be succeeded by the rising generation.

The accomplishments the world will see in the future will come about because of those who live for their dreams, and find a way to bring them into reality. At this launch I met many others including Ryan Kobrick, a Canadian-American earning his degree in aerospace engineering, with dual citizenship and dual purpose in life: to go into space, and go into space again. Heading up the special operations for the X Prize team was Loretta Hidalgo, a young astrobiologist who spends her time traveling the world building cooperation among people to enhance the space industry. Nick Skytland was there, working on his degree from the International Space University, intent on driving progress to make us a space faring civilization. Each of them are normal people, but by working toward their dreams they will reach the achievements they want to see in their lives.

Now this is all fine and dandy while we're looking ahead, dreaming about our future and the life that we're going to live. It reinforces that, yes, we do want to be the rulers of the universe and the masters of destiny. It illustrates that we really can be the geniuses who understand everything, and can develop a solution to any obstacle. It validates that drive inside of us to work long and hard through the challenges and details of our projects. However, we quickly find that creating this grand vision and carrying this unyielding drive is not as easy as watching somebody else do it on television.

Something happens throughout the day, and throughout the careers of many people, which allows that vision and drive to slip away. A price must be paid to reach our dreams, and it is a very real price. It is a price that cannot be measured in dollars and cents. It is a price that must be paid every day, and it is inside of yourself that you find the means to pay that price. Everybody finds his or her source of drive and desire in different ways and for different reasons. We pay the price for our dreams by pulling that desire from within ourselves, and applying it to the cause for which we think and move.

One of the first obstacles to all engineers hits at the beginning of their careers. They step into their company on the first day, armed with all the wisdom and knowledge of an engineering degree, burning with desire to be the smartest engineer this company has seen, ready to put in whatever hours of learning and analysis are necessary to move the project forward ... and then they're given a CD with a date and a couple acronyms scribbled onto it, and told to go crunch data.

"Here. We need this charted. It's a bunch of temperature data from three tests ago. I hope we don't see anything significant in it, but we need to look at it just in case. It'll probably take you a few days, so don't feel bad that you're working on it for a while."

And just like that, the grand vision and feeling of importance vanishes.

"I have to chart old temperature data, hoping to see nothing?"

For some people, the drive hangs on for a few hours as they open the files and start making sense of the hoards of data. But by the end of the day, boredom has set in. Instead of calculating and analyzing with purpose, you're sitting there at your desk fighting off the sense of fatigue that accompanies hopelessness. You're struggling just to stay awake. This continues for the next three days, and every motion of that mouse becomes a mind numbing labor, making your sense of purpose throb in agony.

What happened to the world you hoped for? What happened to the excitement of watching the History Channel as Alan Shepherd became the first American launched into space? What happened to the inspiration from Wernher Von Braun and the engineers who worked night and day to design and build rockets that could hurl men into space? What happened to the enthusiasm from just days ago as you watched with your own eyes while Burt Rutan's team sent up a space ship designed and built in less time than it took you to go through college?

There was so much purpose in all of that accomplishment. There was so much purpose that drove you to grind through college and become an engineer. There was so much you were looking forward to. Why are you now sitting here in this cubicle, allowing your life to be sucked away through a monitor and a mouse, eaten up and lost in the forest of data contained on that CD your manager just handed to you?

Initially, you tell yourself it will just be a passing thing. "Everyone has to pay his dues." Then you're given another CD of data, or even worse, a documentation report to write up. When "paying your dues" looses its motivation, you start trying to justify it with your paycheck. No matter how hard you try, life can never be fully justified with a paycheck.

Where is that purpose for which you started your life and career? The truth is, it is right there in your cubicle and on your computer. It is right there at the company you are working for. It is right there working for the manager and with the team you've been hired to work with.

The accomplishments and dreams for which we live really do exist. We see them accomplished by others on occasion, and look back on them in history. We even reach personal mile markers toward our own larger accomplishments. The truth is we can live the life we want to live, but we must apply some understanding and design if we are to do this.

Engineering requires an understanding of the laws of nature, an understanding of the properties of materials and forces and how they act and react. Engineering requires design, a coordination of elements and forces as components of a system, to behave such that our purposes are accomplished.

Engineering also requires understanding and design of the Human elements. Human elements include the individuals, and the business organization. Throughout your degree, you've learned how materials interact, and what behavior to predict given different conditions. Now, we're going to take a look at how engineering teams interact, and what behavior to predict given the conditions you'll see upon graduation.

Our purpose in this book is to observe several examples and principles that will help you starting off as an entry level engineer. Many of the experiences used herein have come from aerospace, but the principles apply to the automotive industry, commercial products, manufacturing, research, and so forth. All fields of engineering have several common threads, and understanding them can help you be more effective while getting your feet on the ground.

Throughout this book, we'll look at the personal experiences of over two dozen beginning engineers. Each engineer's experience will be used to illustrate a principle, but frequently they could be used to illustrate many principles. The idea is to give you exposure to what these engineers have experienced so that you can learn from their successes, and recognize scenarios and opportunities you may encounter yourself. You will pick up the greatest benefit from this book if you learn the principles addressed, but also draw your own insights.

This book isn't about climbing the corporate ladder, because, quite honestly, not every engineer is interested in climbing the corporate ladder. However, whether you're interested in becoming the Chief Engineer, the Chief Executive Officer, or the Chiefs' Little League Coach, it is important that you get a good start in a good job. This is one step all engineers have in common, so this is the step we will walk through, the step from your degree into industry.

Hopefully, by spending this brief time to generate some realistic expectations of what you may experience, and to understand some of the dynamics of what you will encounter, your first year as an engineer will be more enjoyable for you, more productive for your manager, and more profitable for your company. It will also help you position yourself to move in the career direction that will best suit your personal interest. Engineering can be the best career in the world, and getting started on the right foot can make it that much better.

SECTION 1—GETTING HIRED

CHAPTER 2

Interest and Understanding—*The Substance of Your Resume*

We can't talk about your first year as a new engineer without talking about how you land that first job. We'll walk through what employers are looking for, building understanding (experience), composing your resume, a few extra recruiting tools and methods, and then interview considerations.

Traditionally, you start composing your resume as a one-page fact sheet listing work you have done, and who paid you to do it. However, the resume that will be most effective is much more than a fact sheet, and contains much more than a list of job titles and dates. Your real resume is built on the *interest* and *understanding* you have developed throughout your years in college. You illustrate your interest and understanding using experience.

Experience is the first element people think about when composing a resume. As a result, many new engineers have difficulty composing their resume. Not everybody is able to land that dream internship or get the sought-after technical research assistantship. Many students are working their way through college, and must forego some engineering opportunities for a paycheck. They don't have any experience that feels like *real engineering,* so they feel they don't have anything worthwhile to put on their resume. They may have been part of a team working on a project, but didn't perform any *real engineering* on the project. They feel stuck with a list of experiences that aren't very inspiring, and don't seem like they would be valuable to a company.

The first thing to do is break out of this fact sheet mentality. Begin to think of your resume as a representation of your *interest and understanding,* illustrated through the *experiences* you have gained while pursuing those interests. Your resume is built upon *interest and understanding* because those are the two things for which a company is hiring you.

First we will look at interest, and why it is important. The next focus we will take is to understand what *real engineering* looks like. With an accurate understanding of what *real engineering* looks like, you will be able to effectively illustrate your *interest and understanding* to the company to which you are applying.

Interest

Interest is one of the primary qualities employers look for in new engineers. Interest (often called passion) is of tremendous value to a company. An engineering team wants an engineer who will take a driving interest in the projects they are committed to completing.

If you have had a blazing interest in rockets throughout your four years in college, then you're going to have taken and created many opportunities to work with rockets. If Boeing's rocket propulsion division becomes aware of this, they'll want you on their team.

If you carried a recurring interest in robotics, then you will have taken and created opportunities to work with all kinds of control mechanisms, mechanical instrumentation, electro-mechanical interface hardware, and control algorithms. If Ford's department for automated manufacturing becomes aware of this, they'll want you on their team.

Maybe you have an obsessive interest in grades, honors, and doing well in school. You will end up with a GPA and Honors that reflect your dedication to this.

These cases make good examples, but they're more the exception than the rule. The most common scenario is that engineers are good balanced individuals, talented and driven, but they don't carry a relentless obsession for anything in particular. When asked "What are you passionate about?" they pause, shrug, and say "I like a lot of things. Life as a whole is pretty interesting." Most engineers are interested in many things, and can do very well at any number of them. This is most likely your situation.

The key factor to realize as an undergraduate applying for a job is that employers evaluate your experience in order to gauge your commitment to achieving your objectives. Secondly, they want to see if your interests can be a natural fit with their work. If experience is what they want, they'll hire an expert who has been in the field for many years. If new ideas, passion, and hard work are what they want, they'll be looking for an individual who pursues and works for their interests. They want a new undergraduate with talent and energy.

Applying as a new engineer, you use your experience to illustrate the interests you have chosen, and your diligence in pursuing those interests. In order to illustrate yourself pursuing your interests, you must first have been pursuing them. What that interest is doesn't really matter. What matters is that you have chosen a direction, and you are working in that direction.

Note that an interest is something you *choose*. Maybe some people are born with an innate interest in some particular field, but most people are not. We have an innate capacity for being interested, or being passionate, but we must choose what we apply that interest to.

By choosing where to develop your interest and understanding, you are making a proactive choice as to where you will be in the long run. On the flip side, if you do not choose where to develop your interest and understanding, then you are leaving yourself to the whims of chance. If you think and choose early, it gives you greater control over the direction your life will take. Additionally, it is easier to transition between interests if you have already developed significant understanding in one area. Developing understanding in any field is better than remaining ignorant to them all.

Do something, even if you're not passionate about it! Don't stand still!!!

If you are in your senior year, soon to be graduating, then the majority of your time in college for building interest and understanding has passed. If you didn't get involved in any extracurricular projects during previous years, then what you have is what there is to work with. If this is your case, don't worry about it. You have had plenty of projects in your courses that you can use, you have your senior design project, and you can still get yourself involved with a project at your university.

If you happen to have a couple years left at school, you're in a prime position to develop your interests and build the substance of your resume.

If you are looking for an internship, you will be submitting a resume to apply for the position. Before a company hires you as an intern, they want to be confident that you will take a strong interest in the work they are doing. You illustrate this interest with the experiences you have chosen to pursue before you even contact the company about the internship.

By choosing to be interested in something, and choosing to follow through on that interest, you are building quality substance for your resume. You haven't even written down any dates or project names, and you have begun developing the most important part of your resume.

So that's a little bit about interest, the motivation for your experience. Now what about understanding?

Understanding

Understanding is the most important component you want to illustrate with your resume. This is what companies emphasize most when hiring. Understanding is the result of following through on a chosen interest. Engineering experience is the most effective tool available for developing understanding. But when you're a young engineer in college, maybe you feel like you don't know enough to do any *real engineering*. Maybe you feel that you don't have opportunities to do any *real engineering*. Without any engineering to be done, how can you build any experience? This dilemma is understandable. It's false, but it's understandable.

Opportunities for *real engineering* are all around universities; you just have to learn to recognize them. The trick to recognizing engineering opportunities is in learning what *real engineering* looks like, and where to look for these opportunities.

Real engineering Engineering experience comes from working with a team on a technical project.

Engineering experience will (almost always) coincide with team experience. If you are working as part of a team toward a common goal, then you've met 50% of the requirements for it to be *real engineering*. The other 50% comes when the common goal is a technical project. This may seem like an oversimplification, but that really is all that's required. A team of two or more people, and a project that is technical in nature.

A whole range of things can qualify as a technical project. If you're creating anything mechanical, it's technical. If it has a defined three-dimensional shape, or if it moves, it's technical. If it involves numbers or math, it's technical. If it involves electrons, it's technical.

The level and nature of technicality varies from project to project. This is where you use your discretion about what project you will pick up, and which interest you will choose to apply your time toward. Depending on what interests you choose, the technical details of your experience could look very different.

That addresses what real engineering is, so now let's clear up a common fallacy about what real engineering looks like.

Students often think that *real engineering* looks like homework.

Engineering experience will probably NOT look like engineering homework. Any equations you use will probably be smaller. You may use two or three homework-style equations for your whole project, but your project is still valid if you don't use any. If you use any homework-style equations for your project, keep the paper or program that you used to calculate them (this can be good portfolio material).

When you step into industry, about 5% of your job will look like the homework you've spent the last four years working on. The first place you get exposed to this reality is when you pick up your first engineering project. Your project won't be homework, and neither will your job, so don't be surprised when they don't look like homework.

Do not imply from this that you don't need to know what you are learning through your homework. You need everything you learn, and you need it frequently. However, you usually need it for understanding, instead of for calculating. And for that 5% of the time that you actually put together a set of equations, you cannot get around having the academic background. As an engineer, you may not do much that looks like homework, but learning from your homework is invaluable.

Where to look Learning where to look for opportunities to develop engineering experience is the easier side of gaining understanding. Open your eyes, look around, and spend a little time talking to people you don't know.

Talk to your professors, and talk to upperclassmen. Don't be intimidated. Find out what projects are going on, and choose the one that lights you up the most. People who are involved with interesting engineering projects are usually looking for ambitious recruits to join the team. If they knew you, they would invite you personally. Take the time to talk with them and get involved.

Look at the walls of your university for posters from team projects. Teams usually win awards from competitions, and professors always want these displayed somewhere around the school. Look around for these displays, and follow up on them.

Organizations like the American Society of Mechanical Engineers (ASME), American Institute of Aeronautics and Astronautics (AIAA), Society of Automotive Engineers (SAE), and many others, are always hosting national competitions between universities. Round up a team to enter one of these. If research is your thing, organizations like the National Science Foundation sponsor all kinds of programs that don't require previous experience. If you have something particular in mind and nobody is working on it, then start up the project yourself. Talk with your professors, and they'll be happy to help you get started.

Some engineers start up projects that they do independently, without any team involved. These have value, but their value will increase if you can expand the project to include others. It is difficult to put enough emphasis on the advantage that comes to a project, as engineering experience, when you get other people involved.

The next aspect of where to look for opportunities lies in where you want to build your understanding. Choose a field of interest, or choose several fields of interest, and build a map of engineering skills that those industries need. With this skill map, you can search for opportunities that will give you the understanding necessary for your career interests. You want to focus on *understanding* instead of *experience* because *understanding* is applicable across several fields of engineering. If you're not sure what you're most interested in, don't worry about it. Map the skills needed for possible fields of interest, and start finding opportunities to develop understanding.

Examples

We know what engineering experience will probably not look like, so lets spend some time looking at some real experience that real students have gained to see what it will look like. We will continue with the experiences we introduce here when we move on to the section on composing your resume.

The general sources where students gain experience are Internships, Research, Teams/Clubs, and Class Projects. You have to be hired for some of these opportunities, but that doesn't necessarily make them better. Experience is experience, and employers don't care if you were paid for your work or not. Internships, or summer jobs with engineering content, are very worthwhile, but the scope can sometimes misrepresent your long-term career interest. Club Projects can often become more valuable experiences than a paid position because there is naturally more room for you to choose the technicalities you're interested in. Class projects that you would include come from more specialized courses you have elected to take, so they are very effective for illustrating your interest.

The examples we're going to look at come from engineers with a Club Project, a Class Project, an Internship with a major aerospace company, and an Internship with a small company. Each is typical of the level of experiences available to undergraduate students. Some students create opportunities with more complexity and responsibility, and some build a quality resume with less involved experiences. The critical factor making an experience good for a resume is the understanding you gain. Every path is different, and there are many ways you can create the path you want to lead. The example we are discussing here are the following:

1) Alan Hui: Design–Build–Fly Airplane Competition, AIAA Club Project,
2) Courtney Berglund: Mercedes GLK Design Optimization, Class Project,
3) Bryan Hyatt: Missile Systems, Intern, and
4) Andrew Hosford: Telephone Pole Resonance Program, Intern.

1) Alan Hui: Design–Build–Fly airplane Competition, AIAA Alan was majoring in Aerospace Engineering. He didn't have a "passion" for any particular aspect of aerospace, but he knew some sort of experience would be necessary to land a good job when he graduated. He chose to join a team competing in an aircraft-related competition.

He looked around his university, and the project he chose was an AIAA sponsored, Design–Build–Fly Competition. Teams from universities across the nation would compete in different performance parameters with electric powered aircraft they designed and built from scratch in that school year.

Competitions like this are sponsored by organizations all across the nation, in all kinds of engineering fields. They are readily available to any students who are willing to take the time to search them out, round up a team, and put in the time to compete with other students. In addition to being valuable experience, they always turn out to be a lot of fun. It's hard to go wrong choosing a competition like this to develop experience.

Alan was part of this seven-person team designing and building a 7′ wingspan, carbon composite, electric powered aircraft, specialized for carrying specific payloads and performance measurements. They modeled the design in SolidWorks, built the foam cores wrapped with carbon fiber, and integrated all the electronics and controls components. They used Excel to predict the performance and drive their design with standard flight mechanics equations. The competition was based on how closely the aircraft performed to their predictions for total flight time, energy consumption, etc.

Alan and a partner were primarily responsible for predicting the performance, derived from equations for flight mechanics, fluid dynamics, and stability, pitch, and yaw controls. He and his partner were responsible for these calculations, but they were involved enough to understand what everyone else on the team was doing, and they shared what they had designed with the others. What everybody designed was affected by everybody else's designs, so they all had to collaborate on their design decisions. They each had primary responsibilities, but they all gained valuable exposure to all aspects of the project.

One of the greatest values of this experience is Alan can use it to illustrate an interest in almost anything he wants. He can use this experience to illustrate interest in composites, controls, aerodynamics, structural analysis, CAD modeling, performance modeling and prediction, and performance simulation. If he wants to work for a company with a team in any of these aspects, he can show that he has had his hands on the work, and can speak knowledgeably about why he is interested.

When Alan presents this effectively in his resume, an employer looking over this experience can see several valuable traits, without Alan having to mention a single one of them. They can see Alan working on a team, collaboratively designing several aspects of a project, performing numerical analysis, understanding the properties of composites, understanding the challenges in manufacturing composites, understanding the dynamics of control systems, understanding control command and feedback, presenting the work to reviewers, documenting the work, designing to constraining performance requirements, etc. It's difficult to think of an aspect of *real engineering* that can't be illustrated through this project.

Alan didn't have a passion for small electric powered aircraft. He didn't then, but throughout the project, he started enjoying the small aircraft. He learned to appreciate the challenges and intricacies in this form of engineering. Even more than before, he has an interest in being an effective designer, and in working on challenging projects in aerospace. He chose to be interested in this small aircraft competition, and choosing that small interest has expanded to include more interests.

Experiences like this are available to any engineer at any university. You don't have to be hired to compete in these competitions. You don't have to have any prior experience. These competitions are probably the best way to build your experience and interest, and move yourself toward the career you ultimately want.

2) Courtney Berglund: Mercedes CLK design optimization, class project Many students have a tendency to look down on class projects as second-rate experiences. Some class projects are cheesy, but if you want to illustrate that you're interested in something specific, a relevant class project can be very effective. In addition, if you really apply yourself to a project, making the most of each opportunity, you sometimes find that you really do enjoy the content.

Courtney Berglund chose a focused interest in optimization early on as an undergraduate. She gets a kick out of mathematically modeling everything in a multivariable system, and driving those variables to optimal conditions against set design constraints. It doesn't matter what the system is, she just likes optimizing it.

Courtney was interested in taking a few graduate level courses in optimization, so she got into them as soon as the professors would let her. It was a slight deviation from the typical routine of taking more specialized courses at the end of an undergraduate degree, but she knew where she wanted to go, and it paid off for her.

This class was focused on interfacing optimization software with CAD programs. The idea was to use the optimization engine to control input variables, and leverage the analysis power of the CAD programs for the system model. This way you could build a solid model in your CAD program, configure it for stress analysis, thermal analysis, fluid analysis, or whatever you want, and use the mathematical model generated by the program for your system. You would then plug the variables you wanted to alter into your optimization software, and let the computer churn the numbers until it generated your optimized solution.

As part of the class, they optimized the design of a Mercedes CLK-GTR for aerodynamic flow. She was part of a three-person team on this project, and they successfully took it from start to finish. Having this experience to substantiate her expressed interest in optimization helped her to land an internship with Pratt and Whitney's Jet Engine Optimization team. Experience leads to experience, and she's gone on from there.

3) Bryan Hyatt: Missile systems, intern Bryan Hyatt spent a summer testing rocket valves. When I first talked with him, he didn't think that this internship was *real engineering* experience. This project didn't involve looking up equations learned in class, and it didn't even result in any valuable data about the valves, but it was superb exposure to the nature of *real engineering.* The only problem was, he didn't recognize it.

Bryan was somewhat fortunate when he started his internship for the summer. His manager had a project ready for him to work on as soon as he got there. He was brought into the company's missile division, and they had a potential leakage problem with a valve used on one of their missiles. They needed somebody to test the valves to ensure they performed according to specification, or to find and identify the flaw in the valves.

Initially, this could sound like an exciting position: testing a rocket valve for a high tech missile system (faulty valves make rockets explode), designing and building test equipment, deciding how to take measurements and analyze your data, etc. This is important, high tech work.

Well, these large companies have been testing their hardware since they began generations ago, and for a simple mechanism like a valve, they often have set and defined methods for testing these components. Part of the reason for these defined methods is that from test to test, you need to be able to compare similar data. If you use fundamentally different test configurations for different mechanisms, then comparing the resultant data is like comparing apples to oranges, and it doesn't tell you anything reliable.

Instead of designing new test equipment, Bryan was dropped onto their test floor with their existing equipment, and instructed to learn how to use it, how it works, and what the data mean. This can often be more interesting than designing, because you get to see the genius of several generations of engineers who have worked on performing this one task of measurement. (One of the most surprising things you may find in industry is the complexity, technology, and science required simply for measurement and testing.) Bryan learned the science and engineering behind the test equipment, learned the engineering and specifications for the valves, and started into the testing. Some of the results he got were consistent, and some were anomalous. He tested other valves that had been previously tested and verified as good. He compared data from several tests under controlled conditions.

The results of his findings: Faulty test equipment.

Bryan had worked with his manager and other engineers through this process, so they were aware of the problem he had uncovered. Even though his manager knew of the problem, he still needed to document it. If things aren't documented, they are quickly forgotten, and often need to be redone.

He put together a report documenting what he had discovered, with data supporting why he believed the specific problems were present. He also put together a report including his recommendations for fixing the test equipment. Several of his recommendations were immediately implemented.

By this time, the end of the summer had arrived and Bryan was heading back to school. He never got to finish testing the valves.

The way Bryan's internship is illustrated here, it sounds like he performed some valuable engineering. (It sounds like that because it really was valuable.) However, when I first spoke with Bryan, he felt like the project had been a flop, and no *real engineering* had been accomplished. All he did was find a problem with test equipment. He didn't even find out if the valves he was assigned to analyze were working or not. Universities don't have any classes on "finding faulty test equipment," so it can be easily overlooked as part of *real engineering.*

This example is typical of what many engineers experience, both in the work performed and in their impression of it. The reality is, this is *real engineering*, and we need trained engineers to carry out the job. Additionally, the experience that you gain as an engineer doing something so seemingly inconsequential is essential to making critical engineering decisions later on when you have greater responsibility for a project. It would be easy to overlook the importance of quality testing facilities if you had never dealt with the challenges in designing and maintaining those facilities. It would be easy to misinterpret the testing data you're making design decisions with, if you hadn't experienced how those data are acquired.

For this project, Bryan learned and set up the hardware, learned the software for acquiring the data, analyzed the data in Excel, made a number of charts, calculated a few ratios and numbers using equations from established procedures, thought about the data, ran new tests to verify what he was seeing in the data, came to several conclusions based on the results of his data, and made design recommendations for making the test process work correctly. Is there anything listed there that is not *real engineering*?

This is *real engineering.* The only problem is Bryan wasn't able to recognize it as such, and it was reflected in his resume. First, he had to learn the *value* of his work. We took a detailed look at the work he had done, and it became apparent how necessary it was, and the value it brought to his company. Now that he recognizes the engineering he performed, he is able to illustrate it such that it shows up as an outstanding experience on his resume.

In the section on *Composing Your Resume*, we'll come back to the description Bryan initially had on his resume before he recognized the value of his work, and compare that to the illustration he used after he learned how valuable his work had been. When interviewing, it is astounding how much difference this recognition made in the way he would speak about his experience.

Bryan gained this experience in aerospace, but it could equally be applied to the automotive industry, industrial manufacturing, commercial commodities, etc. Learn to recognize the value of your work, and you will be able to use it to illustrate your value to whatever company you apply.

4) Andrew Hosford: Telephone pole resonance program, intern Andrew Hosford was brought into a smaller company for a summer job. He was just starting his engineering degree and didn't have much experience with any *real engineering*. However, his inexperience didn't matter because he has an engineering mind. Through a series of ideas and accomplishments, he was able to make a very significant engineering contribution to the company.

The company made street utility poles and similar industrial products. When Andrew started, they had him doing some routine tasks that didn't require much of a brain. He had his job, and while doing that, he paid attention to what factors drove the design of these utility poles. It seems like a simple thing, but it's surprising how many variables go into choosing the right utility pole for the right application.

One of the key things they had to calculate for these poles was the resonant frequency. With the wrong resonant frequency, the wrong wind speed could send a street light spontaneously crashing down onto the freeway, or it could drop a traffic light in the middle of an intersection. Traditionally, the field engineers would calculate this for each individual application. This was tedious, and slowed down their response speed for bidding on contracts, but that's the way it was always done.

Andrew learned how they calculated these frequencies, and thought he could write a program that would do this automatically. He had a very small amount of experience programming in high school, but was unafraid to build on the little exposure he had.

He talked with his manager, started into the code, and it wasn't long before he had a working program. He showed it to the field engineers and they liked it. From there he created a user friendly GUI (Graphical User Interface), so the non-computer savvy sales team could use it. This simple program reduced the company's response time to customers from hours to seconds.

This alone was a significant contribution, but one step always leads to another.

Having this analysis tool available, and having a *real engineer* on hand, the company decided to test several of their poles to confirm their calculations were correct. They sent Andrew out to find the instrumentation and data acquisition tools he needed for this. He used a simple setup with an accelerometer and a simple data acquisition program. He tested several poles, and even modified some of the constants used in their calculations until the predictions were accurate within a couple percentage points.

Several years later, when Andrew was long gone, the company was still using his program. They hadn't expected this when he came on for the summer, but that's what he created with his engineering mind. The company didn't *give* him a real engineering experience; Andrew opened his eyes and *created* a real engineering experience.

While gaining experience for your resume, remember that *real engineering* consists of a group effort on a technical project. These opportunities can be found all around your university in the form of club projects, research, class projects, etc. Look for opportunities to be an engineer. Any job with technical surroundings is probably fit for an engineering project. It is up to you to apply your mind to create something useful. Don't downplay the value of your experience. Understand how and why it is valuable, and speak of it in those terms.

CHAPTER 3

Composing Your Resume

We've talked about the substance of your resume, and how it is founded in interest and understanding. Now we'll move on to actually composing your resume.

You have chosen an interest in something, and you have followed through on that interest. Doing so, you have gained understanding, and you have also gained more interest (unless you hated the work, in which case, you have "refined your scope of interest"). Now we're going to use that experience to illustrate your interest and understanding, so that an employer will see the value you can bring to their company.

We'll step through a few basic rules for every resume, the anatomy of your resume and the principles behind each section. Then we'll run an effective recipe for composing your resume.

Rules for Every Resume

Create your resume in Microsoft Word. Don't use WordPerfect, Microsoft Works, or any of those other word processing programs. Every company you'll apply to can use Microsoft Word. If you want to use some other software, and then convert your resume into a PDF, this is also acceptable. Just make sure anybody can open your resume on any computer. If your resume doesn't have a ".doc" or ".pdf" extension, they probably won't be able to open it. If they can't open your resume, it's useless to them, and goes in the trash...along with your opportunity with that company.

A significant assumption is implied in these paragraphs. The day of paper resumes has past. Sure, you'll still bring paper copies to interviews, and bond paper looks nice, but within the company your resume will be circulated through email. Keep this in mind when you email your resume to an employer. It is most likely they will simply forward your email to others in the organization, with their comments on top. Whatever text you place in the body of your email will be carried with your name, acting as a statement of interest, or even a cover letter.

Keep it to a single page.
Be concise and to the point. Don't use 10 words when 5 will do.
Use keywords applicable to the company's technical needs.
Don't get too creative with your resume. You're an engineer, not an advertising graduate.
Your resume is not a place for pictures of your projects. A portfolio has pictures, a resume has words.
No portraits. Don't paste your cheesy smile in the upper corner.

Pick an email address that is professional and intuitive.
Use white, or very light gray or tan paper. Bright green is ugly, and so are other "loud" colors.

You want to stand out and be remembered, but you don't want to be remembered as the "bright green ugly resume guy."

Print and double proofread your resume on a hard copy. Don't rely on the software to catch all your spelling and grammar mistakes.

Always have somebody else smarter than you proofread your resume. (If you think you're the smartest person you know, think again.) It's better to have a friend see your mistakes than an employer.

From here on out, everything is conditional upon your personal preferences. As an engineer, you can take comfort in the fact that the words are 99% of what an employer is looking at in a resume. If the layout or font isn't picture perfect, no big deal. If there isn't "proper balance of black and white space," they won't notice. Put simply, you have to really screw things up for your layout to detract from the words you include. (That being said, you can still make it ugly if you don't think about it. Make sure it's not a train wreck, and you should be fine.)

The Anatomy of a Resume

Header
Objective (recommended)
Education
Engineering Experience
Other Experience (only if pertinent)
Specific Skills (keyword list)
Hobbies and Activities

Header This includes your name, written in big, easy to read letters, and clear contact information.

Give them a foolproof means of contacting you, both now and six months from now. If that means you have two addresses, and multiple phone numbers, then include them. You want to make it as easy as possible for the company to contact you.

If you have a cell phone, including that number and excluding your apartment number might be a good idea. You don't want a roommate inadvertently giving a bad impression to a potential employer. You also want to be sure that any messages they leave won't be forgotten by your roommates, written wrong by your little brother, or anything else catastrophic. If you don't have a cell phone, seriously consider getting one.

Objective If used right, an Objective can be very effective. If used wrong, it can dilute the impact of your resume, or limit your opportunities. It isn't difficult to use an Objective statement correctly, but it must be tailored to the specific company and position. You will have different resumes for different job applications. The Objective will be the most significant change between each of them.

Companies are looking for individuals with a big picture view, and with small picture ability. They want somebody who is interested in the larger project. They also want to be sure you will be diligent in carrying out your personal responsibilities. We'll talk about how you illustrate this in the body of your resume, but for your Objective, you want to express your interest in the main components of the big picture.

Your purpose in an Objective is to explicitly state your interests, without eliminating any opportunities that might be within your interest. This may sound paradoxical, but it is really quite simple.

Example: "Seeking a position in aircraft design, with a primary interest in aerodynamic design, and open to airframe design, structural and thermal analysis, propulsion systems, and flight control systems."

Right there you have told the company you want to be involved in the aerodynamic design, but will be interested in anything else regarding the aircraft. You have expressed an *educated interest* in working on aircraft. Both the Thermal Analysis manager and the Control Systems manager will look further into your resume. They know you are interested, and they know you are at least experienced enough to know the different design areas. They know you have seen the big picture, and looked closely enough to see the engineering needs. They also see you are aware of your interests, so they don't need to worry you will become disappointed with their design field.

We have a good example, so you can recognize when you've done it right. Now, for a bad example. We'll look at each extreme so you can recognize the ways you can hinder yourself.

The most common mistake is in being too general. Some engineers write their Objective using subjective adjectives to the effect of "Self-motivated, creative, and hard working engineer seeking a position in...." These are good traits, but not effective in an Objective. You need to say what is in those clichéd adjectives, but you need to say it in a way that an employer can believe you. Your self-motivation, leadership, and other personal work habits will become apparent in the body of your resume. Clichéd adjectives don't say anything. Concrete interests speak wonders. Your Objective is where you directly state your interest in the work a company does.

If you are really interested in the position to which you are applying, then you will become aware of the various complexities in that field. You will develop an *educated interest* through investigating the field. In your Objective, express your interest in those specific areas in which you could viably work. Instead of only indicating interest in aircraft in general, indicate the aspects of aircraft design in which you are interested: aerodynamics, controls, propulsion, etc. Be just specific enough to demonstrate an *educated interest.*

The other end of the extreme is being too narrow with your Objective. This usually results from an engineer trying to appear decisive, and going overkill.

Example: "Seeking a position in rocket propulsion, designing cryogenic fuel delivery systems for heavy launch vehicles."

This example goes from specific, to more specific. This makes it very limiting. You want to state your primary interest, and open up your options from there. Show focus, and then flexibility.

If you are bent on a position in rocket propulsion and you won't accept anything else, then go ahead and put that in your Objective. If you would like rocket propulsion, but are open to spacecraft, weapon systems, or anything else needing propulsion, then be sure to follow your focused interest with other interests that cover these related areas. Don't worry, an employer won't think you're unfocused if you are open to many interests. However, being too broad can work against you. If you're open to anything, an employer could (justifiably) get suspicious you'll say anything to land a job. Express your varied interests, but be sure they are sincere interests.

In your Objective, you must tailor it to the company and position to which you are applying. If you're applying to General Motors, you don't want to express an interest in aircraft design. Maybe you have two aircraft related projects and an internship with Boeing, but GM doesn't build aircraft. This is where you need to take the universal applicability of your experience, and tailor it to the company to which you're applying. Develop an *educated interest* in the automotive industry, and express that. Write an Objective specific enough that they can believe you really want to work for them.

For those engineers interested in working their way into management, your Objective would be the place to make that clear. However, you do *not* need to state that you're interested in management. Plenty of opportunities will arise for moving into management once you get yourself into the company. Before you can manage engineers, you first must know what it is to be an engineer. Furthermore, companies aren't looking to hire a manager; they're looking to hire an engineer. They will be looking for leadership qualities, but they first want you as an engineer.

In summary, you want your Objective to clearly express your experienced big picture interest. It needs to be relevant to the company. It needs to open, not narrow, your field of possibilities within the company.

Education Put Education at the top. Make it the first section following your heading and Objective.

Employers will take a quick scan through the Education section to answer a couple questions. This is a fact list; nothing more. Make it easy to see the information they want.

Questions they want answered: Major, University, GPA, and Graduation Date.

If you have a minor, list it. If you have any honors, special programs, certificates, or significant achievements, this is the place to include these items.

Try to keep Education to two lines. It's easy to waste space with this section. Two lines should be sufficient.

Indicate your graduation date, or availability date, with something simple and clear. "Graduating May 2007." Even just the month and year is enough. They'll assume you're graduating then. Avoid indications of uncertainty like "Anticipated Graduation." You know you're going to pass your last semester of classes, so plan on it.

GPA is something students are often uncertain about.

Include your GPA. If you don't include your GPA, all they can do is assume the worst. Engineers with a 3.0 include their GPA, so if yours is missing, they'll assume you left it off because it's a 2.5, or something low.

If you have a low GPA (less than 3.0), you may want to leave it off. You'll have to balance that. Somebody will ask about it in an interview. They will want to understand why your GPA is significantly below average. If you worked full time earning your way through school, a note about this is a worthwhile line justifying a poor GPA.

That being said, internet filters often sort candidates according to GPA. How do you get around this? Well, there comes a time to hold your head up and be accountable for your actions. Those with a high GPA have worked to earn that GPA, and the advantages that come with it. They already earned their way through the internet filters, and have earned the right to take pride in their academic performance. It is the reward the top performers have earned for themselves. If you have not earned a good GPA, landing the job you want will be harder. It's not punishment for bad performance. That's just the way it is. The doors aren't closed to you, but you will have to work other routes to get in.

Now regarding the perception about what is a good GPA, and what is not very good. Higher is better, but anything above a 3.0 won't be of concern. A 3.8 may be impressive to an interviewer, but a 3.4 won't detract from his or her opinion of you. If there is a significant difference between your Major and Overall GPA, include both. Open honesty is always appreciated.

What about listing courses you have taken?

In most cases, listing courses is a waste of space. Recruiters know what courses to expect from a typical Mechanical or Electrical Engineering undergraduate. Listing the courses is redundant. If you are particularly interested in a focus one of your courses covered, then you can emphasize that by including a project from the class in the body of your resume. For even more punch, you can take the project a few steps further than the class. You're using your experience to show recruiters what interests you.

If you are graduating in a major that is unusual for the field you want to enter, then a list of unique classes can be helpful. Suzy graduated in Civil Engineering, but she decided she wanted to work on control systems in aerospace, so she took several classes related to control systems. Listing the classes was the only way she could get around the brand of a civil engineering major. It worked.

Some people suggest moving Education down lower in the body. The argument is that your experience is what differentiates you from others, not your education. Their reasoning is correct, but as a new undergraduate coming out of school, employers want to know your major, school, GPA, and graduation date. Putting Education up top lets them clear that question, and move quickly to the substance they really care about.

After you've been in the field a couple years, your GPA becomes unimportant. When applying to a later positions, experience is what really matters.

Another recommendation for after you have been in industry and are looking for your second job is to rebuild your resume from scratch. It will take a couple more hours, but the resultant resume will be a hundred times more effective than simply updating your old resume. If you put in the work, it will pay off.

Engineering experience The section on Engineering Experience is the most important in your resume. Here you will use your experience to illustrate your interest and understanding. Ideally, this will be the only Experience section. This section can include anything from internships, to extracurricular projects, to class projects. Remember, your purpose here is to illustrate your interest and understanding, using your experience, so anything you have done can be used here.

As mentioned earlier, we need to overcome the fallacy among students that internships are "better experience" than club projects, or school projects. Experience is experience. It gives you a platform to get your hands dirty, and get exposure to engineering. It doesn't matter if you have been paid for the experience; all that matters is what you have learned and accomplished.

A further extension on the student fallacy about internships is the impression they must have the formal title of "internship." An internship is a job, somehow involved with *real engineering* (a technical project, where you work with other people). Just because the high council of engineering doesn't sign a referendum declaring you an "intern," doesn't mean your summer job isn't an "internship." Furthermore, it doesn't have to be during the summer; it can be any time or duration, paid or unpaid. Again, understand that experience is experience, and then it becomes obvious that you don't need a formal internship title.

We will focus on this section in detail when we work on composing your resume.

Other experience (non-engineering) This is where you include any non-engineering experiences. The reason you make this a separate section is to put an emphasis on your engineering experience, while holding a reverse-chronological format. Your most recent, or even current, source for earning an income may not be engineering related. Putting this at the top of a lower section allows you to include it in reverse-chronological order, without putting undue emphasis on it. This can be particularly effective if you're still looking for an engineering position after you've graduated.

This is a good place to include effective team and leadership experiences. Team experiences meet half the criteria to qualify as engineering, so they will be valuable.

When illustrating these experiences, use the same principles that we will talk about for the Engineering Experience section. You want employers to see you being an effective employee and diligent worker.

So you were a pest control salesman for the summer? You're not interested in sales, or pest control. But you were responsible for processing orders, keeping track of paperwork, assessing customer needs, and so forth. Don't waste space illustrating the big picture of pest control. The job title is sufficient for that. Jump straight to your performance as a reliable and organized individual.

Don't waste a lot of space with these experiences. You're better off including another class project in your engineering section (to illustrate your interest), rather than another line about a non-engineering job you held for a while. The only reason you include these is to fill in chronological gaps, and illustrate you are a good employee. Remember, you're not providing a fact sheet; you're illustrating your interest.

Specific skills (recommended) Your section on Skills will be a fact list. This will be a list of Key Words for any automatic filtering system to recognize. Include things like programming languages, software packages, CAD programs, manufacturing and machining skills, hardware familiarity, operating systems, foreign languages, etc. Split them into logical categories for ease of reading.

You don't have to be an expert in something to list it here. If you've been exposed to it, enough that you can continue learning it, then include it. If you have particular fluency or expertise with one or another, use an adjective like *Extensive* or *Fluent* before the item, include noteworthy endeavors in parentheses, or some other subtle but intuitive expression of your level of accomplishment.

Hobbies and activities This section is more important than some people give credit.

When we hire an engineer, we're bringing somebody new onto the team, and we're going to be working with them. We'll be working together for 40 hours a week, every week, every month, for several years. We're going to be working on the same project as this new person we're hiring. We're not going to be stuffing this new person off into a corner, and requiring them to do homework all day long. We want to work with cool people. "Cool" is entirely subjective, and many other professions may question an engineer's judgment of what makes a person "cool," but we don't care about that. We're engineers, and what we do is really cool stuff, even if lesser people think otherwise.

Engineers are generally well-rounded people. Often, when you step in for an interview, the engineer interviewing you will share one of the interests you have in your hobbies. Sometimes that's what will break the ice. "So you played racquetball? I played racquetball all the time in college." "You're a diver? Have you ever been out to the reefs off Hanama Bay?" "You spent a month in Thailand? I've never been there. What was it like?" "How was that mentoring program you did for disadvantaged kids?"

Remember you're illustrating your interest in engineering, so don't put undue emphasis on this; you're simply showing that you're a well-rounded person. Also keep in mind that some activities can offend other people. You're probably better off listing "hiking and camping" instead of "killing animals with guns." Political and religious affiliation can be inappropriate unless it demonstrates a skill or leadership experience. Even then, if not treated properly, it can be a turn-off.

You don't need to stress over this section, but include something of your personality here.

Composing Your Resume

The traditional process for composing a resume is to list names of companies you have worked for, positions you have held, responsibilities you have held, and the dates for all this. It is used as a chronological fact sheet. That shows a small picture outline of your history. What this doesn't show is a big picture of your future. You want the company to see your future with them, and you show it to them in the way you illustrate your experiences.

We're going to walk through a simple 5-step recipe for composing your resume. It's not a typical fill in the blanks, generic, bland, ineffective routine. We're going to focus on your experience, instead of your resume. We don't even touch your resume until step 4. It's going to take some thinking on your part. It will take a little time, but it will be worthwhile for both your resume and for the interview you gain.

Your best bet for following this would be to read through the chapter, understanding each step and what you're accomplishing with it. Then bring this book with you to a computer, and do the work for yourself. It will be most effective if you split each step into separate sessions. Work through one step, save your work, and leave it alone for a few hours, or until the next day. Each step builds on the previous, so it flows very naturally.

We'll look at each step and the theory behind it, running through the whole recipe. Then we'll take a look at examples for each of the four engineers introduced earlier. Our 5 steps are the following:

1) Fact List
2) Stream of Conscious Writing and Bulleting
3) Organization
4) Resume Composition
5) Refining and Polishing

Sit down at a computer, open Microsoft Word, and get ready to write.

1) Fact list We start with the quick, easy, and boring part: making a fact list.

Make a list of jobs, club participation, and projects you have been involved with. If you held different positions at the same company, then list those as separate items. If you worked multiple summers at the same company, then list those summers as separate items. If you worked on multiple projects at the same time at a company, list those separately. If you worked on multiple research projects for the same professor, list those as separate items. List any projects you did for classes. List any projects that you have done on your own. Get the idea? We're making a list of everything you have had your mind involved in. You'll be surprised at how long the list becomes.

Again, put down everything. We'll get selective later.

We're not building a resume yet, so don't worry about format. Don't make a header with your contact information. Don't fill in details of what you did. No bulleted descriptions. We're just making a fact list.

For each item, we want all the appropriate information attached. Company name, Professor name, Project name, Club name, Competition name, etc. If a project didn't have a name, make one up.

Include your position title. If you didn't have a title, make up something appropriate. It doesn't have to sound professional. "Company Grunt Worker" is fine if that's what comes to mind. If it was a summer position with a company, and you did some sort of *real engineering*, then go ahead and call it an internship.

List the dates. Month and year for your start and end date are sufficient. If you don't remember for sure, approximate. When employers look at dates, they only look to see the general length and where it fits in your chronology.

After you build your list (or while you're building it), sort things into reverse-chronological order, and separate them into the two categories of Engineering and Non-Engineering. If you're in doubt as to whether or not it was engineering, include it in the engineering section.

By creating this list, we're accomplishing a few things.

Most obvious, we're getting the facts together you will need for your resume.

More importantly, we're collecting a war chest of all the possible experiences and skills you can use for your resume. Students often leave quality experiences off their resume, simply because they never thought of it as resume material.

I was speaking with a graduate student who was hesitant to put his Master's Thesis on his resume because "it wasn't really engineering experience." I almost fell over when he said that. Your Master's Thesis and an Honors Thesis are excellent differentiating experiences, and you should include them (and things like them).

Additionally, we're making you aware of how much engineering experience you really do have. If you think your bag of tricks is empty, but you never look inside of it, then you'll never know otherwise. By doing this, you may be surprised when you prove to yourself that you really have accumulated some significant experience while in college.

If you still have a couple years until graduation, we're making you aware of what experiences you have available. When you learn to recognize experience for what it is, you will be more apt to jump on the opportunities when they become available. You will become aware of what experiences you may need to take your career in the direction you want. You will become aware of what kind of things can be turned into experience. It helps you to start choosing an interest, and begin charting a map to follow through on that interest.

2) Stream of conscious writing and bulleting Now we're going to use a tool, appropriately called *Stream of Conscious Writing.* We'll use this to find out what you did, and more importantly, what you learned, with each of these experiences in your fact list.

Far Eastern religions that emphasize clearing your mind through meditation have pointed out a prominent trait among Western cultures. They say we are plagued with "the monkey mind." Our minds never settle down on one thing. They bounce from thought to thought. The more we try to stop them, the faster they move. In this step, we're going to take advantage of this weakness, and make it a strength. This is not the time to be in a Zen state of mind. Ironically, *Stream of Conscious Writing* is a form of meditation in itself. Go figure.

In this step, we're going to pull everything from your experiences out of your memory bank, and scatter it all over the page. We're going to create a wreckage of notes, statements, project descriptions, technical terms, laymen's terms, and whatever else falls onto the page. It'll look about as organized as a junk yard, and it will be just as useful. Have you ever explored a junk yard? It's a blast. You find all kinds of interesting and useful things. We're going to jump the fence into the junk yard of your mind, and we're going to turn everything over that is lying in there.

Here's how you do it:

(This following section was written in stream of conscience. It's as disorganized as your writing will be.)

- You start writing whatever comes to mind about the thing that it is that you're writing about...your experience, and you don't worry about punctuation or grammar or whatever else your professors have taught you about making lists and whatever else. You write down what you did, probably things will be repeated and redundant but that's ok because you're just letting your monkey mind bounce around in the junkyard where your experience is and take an inventory of everything that is there...but it really isn't an inventory because you're not organizing things into lists, although it could turn out as a list. And you're not doing it systematically. You're just being a monkey playing with whatever it picks up for however long it picks it up and if it goes back to it it'll get back to it. No big deal.
- List out the things you did at your internship.
- Scheduled when I expected to have the frame welded up and we would need to start machining the parts for the heater frames to go in the baking chamber.
- Laid out the schedule on a Gant chart.
- Modeled parts in Solid Works.
- The whole machine was originally designed in IDEAS.
- Created several drawings for machining in AutoCAD because it was quick and easy.

- Used AutoCAD to figure out several dimensions because it was easier than accounting for everything mathematically.
- The machine put freed up seven employees to work elsewhere because it automated what they previously did by hand.
- The tolerances it drilled to were +/−.002" in both depth and position.
- The 80 year old machinist I worked with was one of the coolest guys I've ever met in my life. He worked on everything since WW II and had "war finish" tools that were "black" instead of chrome plated because of the materials rationing that was implemented during the war.
- And if you happen to wander back to another project that you were streaming on earlier and it doesn't apply to the topic that you're writing about at this moment then that's ok too because you can just mouse your cursor back up to that project that came to mind and pick right up with writing about it again.
- And then when your mind is back to what you were on, it will come back.
- Write that you used a MIG Welder.
- Write that you used Gray Iron because it has better bearing properties than other metals.
- Write that you learned about an autoclave for composite parts, and how expensive autoclaves are, so you decided against it, and just went with a wet lay up, which worked fine for your needs, but you know something about autoclaves even though you didn't use it for your project.
- And if any editor will let me keep this stream in this format and publish it . . . I think I might be surprised.
- But that's the way you let your mind roll.
- We'll look at more examples.
- Remember the technical words for what you did.
- Keep writing and describing the technical work or instrument using layman's terms until it comes to mind and you can write it.
- That is important because technical words are valuable.
- And if you have to Google something to find a technical word in the middle of it, go right ahead, but be sure you move quickly and get right back to your writing as soon as you can.
- And we can't forget about the writers block thing that so easily sets into people.
- Actually, this is a tool to break through writers block.
- But what you do is just write and write and write and keep on writing even if your mind hits nothing other than the words "I don't know what to write I don't know what to write I don't know what to write I don't know. . . ."
- Just keep on writing that because eventually you'll get bored of writing "I don't know what to..." and something will come to mind about your experience and then you can jot that down without the slightest hesitation and get right back to writing "I don't know what to write" because you have a monkey mind and if you try to write the same thing you'll get flashbacks of spelling quizzes in third grade and you'll feel like your being punished and your mind will already have drifted to another thought and hopefully that will be about your experience because that's what you're distracting your mind from and that is exactly what you're looking for.

That's what a mental junkyard looks like, and that's how you use your monkey mind to explore it. It gets really long, really fast.

For each experience, you will create two different thought streams:

1) the Big Picture of the project, and
2) the Little Picture of your activities.

We're trying to pull out two things from this exercise: your interests and your understanding. The end purpose is to communicate both to employers. We will use the Big Picture to illustrate your interest, your overall engineering understanding, as well as your sense of being a team player. We will use the Little Picture to illustrate your reliability, appreciation for details, and ability for technical familiarity.

While streaming on the Big Picture, focus your mind on the overall project. What was the purpose for the project? What were the goals of the project? To which kind of industry does the project belong? What fields of technology were used on the project? How was this project valuable to your employer? How could this project be valuable to anybody?

You are exploring the entire project itself, and all its far reaching applications. Don't think about what you did or did not work on; we're just looking at the project. It doesn't matter if you didn't do anything that contributed to the electrical guidance systems of a rocket you worked on, write an explanation of how the guidance systems functioned. Write about the propulsion system (which you worked on). Write about the recovery system, even though you had nothing to do with it.

The Big Picture of a project is the aspect most commonly missed by engineers while composing their resume. This aspect also has the greatest influence on the apparent significance of your individual work. By writing about all these aspects of the Big Picture, you are illustrating the background and the setting of the work for which you were responsible.

Employers are looking for engineers with a sense of the Big Picture. They want an engineer who is aware of the project as a whole, and who will grow into roles of increasing responsibility. It is worthwhile for you to take the time to become aware of the Big Picture of projects you have already worked on.

While streaming on the Little Picture, focus your mind on what you personally did. Include all the equipment you worked with, the software you worked with, how many hours you spent on something particular, etc. If you fell asleep twice because you became deathly bored reading a product manual, go ahead and include that. But include a description of the product manual, and how it played into what you did to make your project work.

By the time you're done with this streaming, you want to have used every technical word you used for working on this project. Servo motors, drivers, PCIB cards, FEA, thermocouples, diodes, script files, and so on. You want your full technical vocabulary associated with this project written in here. If they don't naturally come up, then keep writing words while you are searching for the technical words. Whatever it takes, get them written down.

Don't get hung up deciding if something is Big or Little Picture. Just write it wherever it comes to mind.

Since the stone cutter analogy is so useful, we'll use it here even though you've heard it before.

A man walks down the road and sees two men cutting stone with exactly the same method. He asks the first what he is doing, and he replies, "I'm cutting stone." He asks the second what he is doing, and he steps back, gesturing to all the construction nearby, and replies, "I'm building a cathedral."

This is the way you need to look at the engineering you are doing. Employers wants to hire somebody to help them build their cathedral. They need somebody who can see both the cathedral, and see the stone.

Run this stream of thought routine for each experience in your fact list. Spend as much time as feels right for each experience. Spend more time on the engineering-related experiences. Chances are that your mind will drift back and forth between experiences. Feel free to move on to another project when thoughts for the one you are on seem to run dry.

You do this for several reasons.

Even though it is highly disorganized, it is an organizational tool. We're getting everything you have to work with out on the table. It's easier to eliminate words than to create them, so we're doing a stream of mass creation here.

It's like creating a billet of aluminum that you will later take to a machine shop, and use your mill and lathe to cut it down into a useful part. First you need a chunk of metal to start with. You will cut most of it away, but having that big chunk at the start is essential so that you have excess material to eliminate. Your finished product will have far less mass than your original piece, but it will be a precise and valuable tool when you're done.

In the next step, you will organize and eliminate words from this streaming mass you're creating, but you will find the important words you want to keep, and that will become your resume.

The second reason we're doing this is to bring both your experience and interests back into your conscious thought.

You want to become conversant in what you have done, and what you are interested in doing. You can't talk about something you don't remember. During an interview is not the ideal time to be jogging your memory for an experience that happened three years ago. Now is the time to bring all those memories to the forefront.

You're remembering and recognizing what you've done for the purpose of interacting with other engineers. In a way, this is like studying before a final exam. You vaguely remember everything you learned through the semester, but it flows more easily if you spend a little time reviewing everything you covered. You could figure it out on the spot if necessary, but it would take three times as long. When you are in an interview situation, you will want to have your interests and experiences fresh on your mind so you can fluently communicate about your technical interests, illustrating them with your technical experience.

3) Organization (your experience document) At this point, you have a whole smattering of run-on sentences and disorganized bullets underneath each experience. Now that you've put down everything under the sun relating to these experiences, it's time to start organizing.

Here's where we begin consolidating your streaming thoughts. We're not putting together your resume yet; all we're doing is organizing the thoughts written down. We're still keeping everything under each heading you created in your fact list, and in their Big and Little Picture categories.

You want to clump the thoughts, skills, tasks, technologies, terminology, etc., into logical groups. Bullets are good for this. If you like paragraphs, those can work. You may want to begin by cutting and pasting, and dragging and dropping phrases into similar bullets. You will want groups for programming, software skills, CAD programs, hardware skills, manufacturing skills, etc. Whatever logical groups appear on your paper will work.

Here you will also consolidate from unnecessary details to relevant details. Nobody cares that you had 14 temperature sensors, 3 optical sensors, and 4 magnetic position sensors. However, we do care that your control system had 21 inputs for the feedback algorithm, comprising temperature, optical, and magnetic position sensors. In addition, we care that your control algorithm was programmed in C++.

Redundancies and duplications of parts of each experience will probably exist. You don't necessarily need to eliminate these, but you need to make sense of each bullet. Eliminate the stuff that is irrelevant, and make sense out of the rest.

A useful way to approach this section is to treat it like your portfolio. Organize it as a collection of more in-depth descriptions of each of your experiences. Organize everything so that it makes sense to other people, and covers the technical aspects of your projects.

It is a good idea to keep this completed section around, and to refer to it whenever you are tailoring your resume for a different company. It will also be helpful when you are rebuilding your resume to seek out your second job in industry.

The only real difference between this and your portfolio is that your portfolio will include pictures, diagrams, and attached project reports. We will discuss your portfolio a little more in the respective section.

Save this document as it is. (You have been saving all along, haven't you?) This is the source from which you will build several resumes, according to where you're applying. From here, we'll refer to this as your *Experience Document.*

4) Resume composition This is where the real work starts, or rather, this is where building your resume becomes an art. Here you will really begin picking and choosing what, how, and why you use the words that you choose.

First, sit down at a computer and open a new Microsoft Word document.

Write out the information for your header. Name and contact information. Don't make it look good yet. That'll just distract you. Leave it plain.

4a—Choose an objective Choose the company to which you want to apply. Maybe you're tempted to think, "I'm just making a general resume right now." Choose a company and a field anyway. It won't hurt you.

Now that you have a company and career in mind, you will begin tailoring your resume.

To begin tailoring your resume to a position at a particular company, start off by asking yourself: "What do they do that I can be interested in?"

Because every engineering project has several aspects, and every business has several projects, you can generate several answers to this question. If you want to do some stream of thought writing on the answers to this question, it probably wouldn't hurt. More clarity is always better. With the internet, and the effort companies make to match the right person with the right job before they even apply, ample opportunities are available to find out what a company does before you send them your resume. Networking is the best way to learn. Find out what the company does and take an interest in it.

You are not asking, "What would they want me to do there?" You don't know enough about what they need new engineers for to answer that question. Also, that is a Little Picture question. It will lead you nowhere. Look at the Big Picture of the company, and find what interests you. They will assign you to a Little Picture responsibility, but they want engineers who see themselves working on the Big Picture.

The interests you find in this company are what you will tailor your resume toward. You will use your experiences to illustrate your understanding that can be useful to the projects the company is working on. The most effective way to do this will be to start by looking over the Experience Document you have created. That document is probably a few pages long, so you already know you won't need "filler material" in your resume.

From your Experience Document, evaluate each experience you have from the perspective of relevance to the interests you have chosen in the company to which you are applying. Don't worry about what experiences are most impressive. Evaluate which are most relevant, and mentally rank them.

4b—Copy over experiences Now we're going to start filling in your resume.

The best way to start may be to copy the bulleted experiences you choose to include directly from the Experience Document you have created. Start with your

Engineering Experience. We'll only bring in your Non-Engineering Experience if it helps your message.

You will keep the experiences in reverse-chronological order, but you potentially have some leeway on this. If you had simultaneous projects, you can place the most relevant experience first. If you are still involved in multiple projects (and you should be), you can choose which to place first. If it is a class project, you can place it chronologically, or toward the bottom.

Your senior design project is as viable and valid as any internships or club projects. You can place that at the top. But if it is less relevant, you can place it near the bottom with other class projects.

Your employer won't scrutinize your resume and its organization nearly as much as you do. Just make sure you don't have any glaring inconsistencies and you should be fine.

Again, don't make this pretty. Just keep it as an outline so you don't distract yourself with fluff.

4c—Consolidate Once you have organized the experiences you want to use, you will begin the final wave of consolidation. What you pasted in will be too long for the single page, so you're going to have to remove the least important elements and combine the most important. But before you start combining and reducing, let's take a look at what you want the end result to entail.

You are going to illustrate two things with each experience. First is the Big Picture of your project or position. You want to give the interviewer the setting in which you worked. (Since we didn't mention it, your interview begins the moment the company opens your resume.) Next, you will move onto the Little Picture, wherein you will illustrate your ability for technical understanding.

The final format we're going to have for each experience will be the Subject Line (fact list), and two to five bullets.

The first bullet will be the Big Picture. That's all you get (maybe two bullets). You are limited to being concise, so be sure to make it comprehensive. You are painting the background in which your individual activities will be seen. Illustrate the field and technical nature of the project. State the goals, purposes, or function of the project. Whatever you say, be sure your reader can see the Big Picture of what you and your team (or company) are working on.

The next one to four bullets will cover two things: your individual activities and the technical challenges you personally addressed.

The technical challenges are more interesting to read and write about, and they are the elements that illustrate your technical aptitude. The activities you performed to address these challenges were done for just that purpose, to address the technical challenges.

Coincidentally, the most effective way to illustrate this combination is to state that you performed X activity to solve A, B, and C technical challenges.

Maybe you created a solid model using CATIA to integrate the power, control, and data collection mechanisms, within competition design constraints, minimizing exposed surface area, while allowing for convective cooling. That says a lot more than "Solid Modeling using CATIA." It shows an engineer at work, solving multiple and conflicting design constraints.

The bottom line of what you're trying to do with these bullets is to show the employer that you understand the multiple variables involved in your design. You are showing them that you understand your project inside and out. You are showing them that you have a full range engineering mind.

Be aware of how you use acronyms. An interviewer will know what FEA means in a bullet about structural design, but won't have a clue about what the UCAV team is (unless identified as Unmanned Composite Air Vehicle). Even worse, the interviewer could interpret the acronym as something else. If in doubt, spell it out.

You will reduce each of your experiences into bullets in this manner, but as you do so, you will keep in mind what skills are already present in your resume. It is likely the design skills you used in one project were also used in another. It is also likely each project was unique, and as such, required different design considerations and activities. With that in mind, you may want to emphasize a particular skill set with one project, and emphasize another with the other project. You will identify that you used both skills in each project, then you will place a different emphasis in each. This can illustrate a more comprehensive field of understanding.

We could walk through many other skills for composing your resume, but if you keep these principles in mind while writing, these methods will become apparent to you:

1) You already have plenty of experience; you just have to put it into words.
2) You are expressing your interest in the Big Picture, and your skill in the Little Picture.
3) You are using your experiences to illustrate your technical ability and understanding.

After you have composed the body of your resume, it's time for the finishing components: Objective, Education, Skills, Hobbies and Activities. You put these in last because you don't want to crowd out the space for your engineering experiences, which are the most important. We covered these earlier, so fill them into their appropriate slots.

Here's a quick refresher:

Objective: Express an *educated interest* that opens your options within the company.
Education: Two lines; include your Major, School, GPA, and the Month/Year you will graduate.
Skills: A Key Word search list.
Hobbies and Activities: Evidence that you're a rounded person, and interview ice breakers.

5) Refining and polishing This is the least important step in developing your whole resume. If you're split between polishing your resume and going mountain biking, you probably should go mountain biking.

We've already brought your experience into words, and composed those words to illustrate your interest and understanding. That is by far the most important part. As we covered earlier, you really have to screw things up for the layout to detract from the words you include. That's not to say you can't screw up. Slop and confusion detracts from your words. Make it look good, but just don't stress about it.

Now we'll look at a few considerations for polishing.

Make your resume fit onto one page. If you're a graduate student or applying to your second job, two pages are fine (if necessary). You may have to tweak your fonts and margins, but that's not hard to do. Everyone has preferences about what looks best. Find something good enough for your taste, and go with it.

Header: Put your name in the top center or upper right corner. This makes it easier for recruiters shuffling through a stack and finding you. Don't waste a lot of space with your contact information, but make sure it's big enough for the HR office (Human Resources) to read without squinting.

Choose a font that will not take up a lot of space. Wide fonts can cause a bullet item to wrap to a second line after only a few words. Sans serif fonts (ones without small lines at the ends of characters) usually take up less space and are easier to read. Also, choose a common font, such as Arial or Helvetica. Not all computers have all fonts. If you use a rare font to look distinctive, and a person viewing the resume does not have that particular font, your resume will loose its format on his or her computer.

Use the same font through the body of your resume. Switching fonts between subjects and bullets may make things difficult when you make revisions. The "visual advantage" you gain from a clever combination of fonts can be gained by using different font formatting: bold, italic, and font size.

Save your resume with a simple name that is distinguishing, yet intuitive. "LastName_Resume.doc" works well. Avoid "Professional Resume.doc".

The next time we're looking for "That Andy guy. . .Andy Rie. . .Rie-something. . . . Where's his resume?"

If you don't have your first or last name on the name of the document, we might loose interest before we find you.

"I'm not finding it. But here's that Allison girl we brought through. She was sharp, and picked right up on. . . ." Andy just lost the game.

Examples

Now we're going to look at how the four engineering students we introduced earlier developed their resumes to represent their interest and understanding. With each student, we worked through their resumes during different Resume Workshops. In the workshops they already had their current resumes, so we built from there. We'll look at before and after, as well as in process examples.

1) Alan Hui: Design-Build-Fly Airplane competition, AIAA When we first pulled up Alan's resume in the workshop, this is how he represented his experience with the AIAA Design-Build-Fly Competition.

AIAA PROJECT, AIAA STUDENT GROUP, UCLA **PRESENT**

- ❑ Participating in building a remote control aircraft to compete in national tournament.

We began talking about the details of the project, and typing in bullets. These bullets are listed just as we jotted them on the spot.

AIAA PROJECT, AIAA STUDENT GROUP, UCLA **PRESENT**

- ❑ Participating in building a 7' remote control aircraft to compete in national tournament.
- ❑ DBF (Design-Build-Fly) Take project from scratch.
- ❑ Design in SolidWorks
- ❑ Written Report on Design dimensions,
- ❑ Predicted performance using aerodynamic modeling equations, basic flight mechanics, computational modeling in Excel
- ❑ Wingspan 7'
- ❑ Electric motor, flys for 10 minutes
- ❑ Can be disassembled and stored in a 2' × 7' × 4' transportable
- ❑ Mission: hold two external payloads on the wingtips, and one payload in the fuselage. Payloads are 3 lbs each.
- ❑ 7 person team. Competition in Maryland
- ❑ Responsible with a partner for calculations for predicting performance
- ❑ Calculations involving fluid dynamics, stability, pitch and yaw controls
- ❑ Built from carbon fiber, Designed mold, wrapped carbon sheets around the mold, and applied resin. Cut the aircraft out of the mold and
- ❑ Foam core with carbon wrap for wings and tail
- ❑ Servo motors controlling stabilizers and ailerons

Alan took those notes away from the workshop, and consolidated them for his final resume. You'll notice how Alan consolidates all the details into logical groups.

AIAA PROJECT, DESIGN/BUILD/ FLY COMPETITION, UCLA **2005**

- Participated in designing and building from scratch a 7' wingspan, electric powered, remote control aircraft with a team of 7 members for AIAA Design/Build/Fly competition.
- Airplane could carry and remotely release 3 lbs payload from each wingtip, be disassembled and stored into a 2' × 7' × 4' transportable box in 30 seconds.
- Designed in SolidWorks, and built aircraft body and components with carbon fiber, cut molds from hard foam, and used various machining tools including water jet machine.
- Collaborated with a partner to predict aircraft aerodynamics performance, longitudinal stability and pitch control using Excel.
- Wrote a design report on aircraft design details, performance, and dimensions.

If your resume isn't full, it's not because you lack experience, it's only because you lack detail. As a case illustration of the fact that you have plenty of experience to fill your resume, here's a clip from the email Alan sent with his finalized resume:

> Because I have included more details for my projects, I had to exclude my employment experience section and some other projects I have done due to space limitation. Is this a good thing?

Yes, this is an excellent thing.

Everything that Alan now has on his resume is good quality engineering substance. We still wanted to include a couple of his older engineering projects, so we shrunk his margins, consolidated a couple more lines in his experiences, and cut his Education section down to one line. We inserted two other projects, made sure the resume didn't look crowded, and he went forward with that.

2) Courtney Berglund: Mercedes GLK design optimization, class project For Courtney's class project, she took all the project goals, design tools, and system concepts, and using the technical terminology, merged them into two bullets representing the big picture and the little picture. We use this example because it is a highly effective demonstration of how to create a resume-style run-on sentence, illustrating the technicalities involved in a project. Note her use of the technical terminology and engineering tools.

MERCEDES CLK GTR DESIGN PROJECT — **SEPT 2003 – MAY 2004**

Brigham Young University, Provo, UT

- Part of a three-person team to model a complete to scale Mercedes CLK-GTR using CATIA for aerodynamic flow optimization.
- Created a cad-centric tool-kit to analyze and optimize aerodynamic flow using CFD methods in Hypermesh and Fluent and optimization algorithms and processes in iSIGHT to iterate and define an optimal design within the design space of the mathematical model.

3) Bryan Hyatt: Missile systems, intern Bryan initially had his experience poorly represented with three bullets.

RAYTHEON, INTERNSHIP — **2003**

- Tested valves for sporadic gas leaks
- Successfully completed the Fundamentals of Tactical Missile Design Course
- Created and revised CAD drawings in ProE

What's worse, when we talked about his experience, the first thing that he said was, "It wasn't anything useful. All I really did was try to make broken test equipment work." He wasn't aware of the value of what he had done. We talked about what we mean by *real engineering*, and it turned his perception of his experience upside down. It hadn't occurred to him that test equipment is designed and improved by engineers.

Look over his revised resume, and then we're going to point out several comparisons between the two. We've also included his illustration of his Senior Design Project for

RAYTHEON MISSILE SYSTEMS, AIM-9X MISSILE TEAM — **2003**

Engineering Intern—Analyzed cause of sporadic gas leaks in missile guidance unit

- Evaluated design specifications and hardware conformance to specifications for both in-house and vendor-supplied missile components
- Designed and executed testing that exposed faulty testing equipment on the production floor
- Documented and presented results and recommendations to missile team leadership, including improvements in design, assembly, and inspection, some of which were immediately implemented
- Successfully completed the AIAA Fundamentals of Tactical Missile Design Course
- Created and revised CAD drawings for airframe components in ProE

a few things we'll point out. His attitude about his Senior Design Project was equally dismal the first time we spoke.

ERIC TAYLOR RESEARCH: CAPSTONE SENIOR DESIGN PROJECT **2003 - PRESENT**

Student Engineer-Designing filtering device for protein isolation process in biomedical research. Design improves upon existing disposable technology by consolidating multiple components into a reusable, self-contained system, and adds interchangeable filter sizes.

- Creating functional prototype to be used for demonstration and marketing
- Primarily responsible for project planning, scheduling, and management using MS Project

The first difference you notice between the two is the detail Bryan included in the second illustration. We can actually see what he was doing. We can see he was busy working with hardware, performance specifications, designing and executing testing, working with the test equipment and understanding it enough that he could tell it was broken. He analyzed the data the equipment was giving him, and he understood it. He didn't just run through the motions and make charts. We see him effectively communicating with his management team and reporting on his project. We see him creating solutions that were immediately implemented. We actually see the *experience* Bryan had during that summer internship.

Two more subtle differences exist between the two. They don't immediately stand out, but they make all the difference in seeing the full experience. At the start, he paints the Big Picture, and follows with the Little Picture.

In the project heading, he identifies that he was working on the AIM 9X Missile Team. Not just a summer intern at a big name company. Now any Missile system team will immediately start looking at him.

Just below the header, he paints the Big Picture he was working on. He identifies he was an intern, and uses that to show what his role was for being on the Missile team. Immediately, we begin to see a picture of Bryan working toward a defined purpose.

For his Senior Design Project, he uses that same method to illustrate the Big Picture of the project his team is working on. Immediately we see this is a high tech device, and he is part of a team creating a valuable product that improves over existing designs. We see him as a successful engineer, but also as a conscientious contributor to the overall business. We're looking at the beginnings of engineering leadership.

A few months later I was at his university interviewing new engineers, and I was close enough to overhear while a coworker interviewed Bryan. I was floored at the difference in how he talked about his internship and his Senior Design Project. It was night and day from how he had been talking about them just months earlier. His conversation went from downplaying an unimportant project, to describing technical details and challenges of the work he had participated in. Everything he said was consistent with what we had talked about originally. He wasn't just putting on a smiling face. He knew his work was important, he knew it was valuable engineering, and his interview brought out the understanding and expertise he had developed.

Ultimately, Bryan decided he wanted to go with missiles. We do satellites, so he took a missile opportunity, and we couldn't get him on our team.

4) Andrew Hosford: Telephone pole resonance program, intern Andrew was majoring in Mechanical Engineering, and he was brought on for the summer with a smaller company in his area. As we saw earlier, Andrew turned this into an excellent experience that was a tremendous benefit to both the company and himself. Here's how it initially appeared on his resume.

SOFTWARE DEVELOPER

Shakespeare Composite Structures Summer 2002 Newberry, SC

- Developed and beta tested detailed software that accelerated company response to customers from several hours to seconds.
- Diagnosed software errors and developed solutions to resolve software problems.

It doesn't sound bad. It actually sounds pretty good, but it doesn't nearly do justice to the real work he did. It certainly doesn't illustrate the skill he developed with programming, numerical analysis, data acquisition, and equipment testing. It doesn't show the initiative demonstrated either. We pulled up the keyboard and started writing.

STRUCTURAL ANALYSIS SOFTWARE ENGINEER

Shakespeare Composite Structures Summer 2002 Newberry, SC

- Developed and beta tested detailed software that accelerated company response to customers from several hours to seconds.
- Diagnosed software errors.
- Used a windows based platform to create program.
- Generated iterative method to develop numerical solution to discovering the natural frequency of any given company pole.
- Used Building codes to determine and iterative solutions to develop software that would determine if a pole would withstand any given wind load.
- Generated a mathematical model to determine iterative bending moments of finite pole segments to determine final bending moment, under variable conditions and variable distributed and fixed loads.
- Application automatically calculated moment on variable and dynamic pole section.

- Fiberglass light pole.
- Analyzed winding code (used for automated manufacturing) process to determine regions of variable thickness on pole.
- Program was verified using actual test data, determined by placing an accelerometer on tip and verified that projected natural frequency was the same as real natural frequency.
- Iterative process of mechanical comparison to projected data, then adjusting program to be more accurate.
- Presented data to Board of Trustees and received positive feedback.
- Created user friendly GUI.
- Application is still in use after 2.5 years with minor changes or improvements.
- Developed solutions to resolve software . . . .

One thing we don't have included here is the programming language. We're also missing the equipment he used to test the poles. And finally, we didn't write much about the Big Picture. But this is the right idea. It shows how much more involved projects are than typically represented on a resume. Familiarity with test and data acquisition equipment is very valuable, but it's easy to overlook because it's not the actual product of focus.

It is significant to note that he changed the title of the position from *Software Developer* to *Structural Analysis Software Engineer.* This is one of the ways you express your interest. Andrew doesn't want to do general software development; he wants to develop software for engineering analysis. This title paints the picture of Andrew being a mechanical engineer who is fluently skilled with computer analysis of structural systems.

Companies will want to hire him to do more of this type of work. He's not misrepresenting his experience in the least. Actually, changing that title is more accurately representative

STRUCTURAL ANALYSIS SOFTWARE ENGINEER

Shakespeare Composite Structures Summer 2002 Newberry, SC

- Developed engineering analysis software for use by field engineers and sales team for structural fiberglass poles accelerating company response to customers from several hours to seconds.
- Generated a mathematical model to determine the natural frequency of poles with variable cross sections under dynamic conditions with distributed and fixed loads.
- Designed and executed testing with accelerometers to verify calculated accuracy within +/−4%
- Analyzed automated manufacturing process to model regions of variable thickness on poles
- Presented the program and test results to Board of Trustees
- Application is still in use after 3 years with minor changes or improvements

of his experience. The important principle to learn is that with subtleties like this, you can show yourself doing what it is you are interested in doing.

Other Resume Considerations

Inflating your experience One hang-up some students arrive at is that their inherent modesty sets in when they are illustrating their experiences. In their perception, their experience doesn't match up to the Hollywood version of *real engineering*, so they don't feel it is very impressive. Since when was Hollywood real? Then, when they accurately illustrate the work they did, and do so in a way that makes it appear important, they fear they are inflating their experience. Actually, they're not inflating anything. They're just seeing the value of their work as they've never seen it before.

That being said, you still need to be cautious about inflation. Be sure you're not misrepresenting anything. A good gauge to check with is putting yourself in the position of your manager or of a fellow team member. If they read what you wrote, would they disagree with any of it? The best thing you can do is actually send them your resume and ask them if they feel you inflated anything. You can even ask them for comments at the same time. They will be happy to see your quality resume. Teammates will be glad you trust them enough to let them see your resume. They'll probably want to borrow the way you illustrated part of your experience. And finally, any comments you get back will only be useful.

If you can deliver on the claim, you're not inflating anything.

"Fluent in C++." Are you? Would an interviewer be disappointed if he put you on the spot? That is the question you really need to answer. Maybe you haven't used C++ in an internship, but you've been tinkering with it since high school and can program like a ninja. Sure, you can put that on your resume. If you're that good though, and you enjoy programming, you probably have some experience you want to put in the body of your resume. If you don't have a project using your skills that can fit on a resume, then create an opportunity or team project to use your skills. Doing this will also stack up as points for leadership. Find a way to put your interest into experience, and from experience into words.

Proprietary experience Some engineers have resumes listing the company, location, and dates. Then in place of a project description they simply put "Proprietary work with non-disclosure agreements."

We can do better than that.

If you happen to be working on a proprietary project and have signed a non-disclosure agreement (because they'll be patenting your project), this won't be a problem. You don't want to compromise the proprietary work you're doing, but there is still much you can do to illustrate the experience.

What can you put on your resume from your proprietary experience?

Let's answer this question by looking back to the principles you build your resume on, and the format you use to illustrate your engineering interest and understanding:

1) You build your resume to illustrate your interest and understanding.
2) You illustrate each experience with a bullet on the Big Picture of your project, followed by bullets on the Little Picture details of the project.

We have these two design objectives, with a design constraint of avoiding any proprietary disclosure. We can engineer your words to make this work.

You already wrote about your project, just like it was any other project. Go ahead and format that for your resume like you would if there weren't any disclosure constraints.

Now looking at what you've put together, ask yourself: Is there anything in those few bullets that discloses the novelty of the project in any more detail than what is already common knowledge to somebody in the industry? Probably not. If there is, then all you need to do is dilute it until it is general enough to not disclose what is proprietary.

The Big Picture bullet gives a description of the function and purpose of the project you are working on. You were working on a better mousetrap, which will trap a larger range of big and small mice, do it more reliably, and use less cheese. That illustrates the Big Picture of a mousetrap without disclosing anything proprietary.

The Little Picture bullets give a description of the engineering details you were involved with. You used accelerometers to measure the resonant frequency of the mousetrap, programmed the data acquisition system, analyzed the data in MATLAB®, created a solid model of the mousetrap in Pro-E, and used an EDM for precision machining on the parts. Any proprietary disclosure in that? Nope. But there's a lot of engineering. Given the amount of space available on your resume, you'll really have to work to disclose anything in such a way that forfeits any patent rights.

Again, what you include in your resume is a representation of interest and understanding, not the project design. If the highlight of the project is that it was proprietary, then the interest that is conveyed is "I'm interested in working on proprietary projects." However, if the highlight is the electromechanical interface, it portrays a much more useful interest.

While composing your resume, remember that your purpose is to express your interest and understanding, using your experience. Employers are looking for evidence that you are interested in their work, and that you follow through on that interest. Use the Objective in your resume to express an educated interest, without limiting the scope of your interests. Organize your resume with engineering experience first, to put the emphasis on the most important aspect. Use your monkey mind to put your experience into words, and then consolidate it to concisely represent your interests, skills, and understanding. Use all the key words that are appropriate. Illustrate the Big Picture of your project, as well as your Little Picture activities. Include which skills you used to reach which design criteria, or which functional constraints were considered to determine which design decisions. You don't want to inflate your experience; you want to represent it articulately.

CHAPTER 4

Tech Team Tryouts—*An Interviewing Overview*

Thousands of engineers show up for thousands of interviews every year. They all make it through one way or another, and never get any more coaching than what they discuss with their friends over homework. It's not terribly difficult. You don't know what to expect, but you'll make it through. Since a few common patterns can generally be anticipated, we'll look over these.

We're going to start off with a few notes on applying. Then we'll move into what to expect in an interview, and look at some actual interview scenarios. Most of the theory behind how to run an interview will become apparent in the scenarios. We'll also address how to prepare effectively for interviewing, and look at a few other considerations worth keeping in mind.

Applying

Applying for a job is the connecting link between your resume and interviewing. You can go about applying a few different ways. You can take the mother hen approach, carefully preparing and watching each application, or you can go to the other extreme with the shotgun approach, thoughtlessly sending your resume to anyone and everyone. Both methods can work, but neither extreme is fail proof.

Your University Career Services is one of the most important places to work through. In order to use career services, you must be registered. Make this part of your first week of the semester. Get yourself registered and upload your resume. More often than anybody would like, companies are disappointed with the number of candidates who have applied through the career services. Many quality candidates miss opportunities for good jobs because they're one week late getting their resume into the system. It's not because they're irresponsible. It's because they aren't aware of the deadlines.

Consider this your official notice. Companies recruit through your University Career Services department. Get on it!

Web applications are effective. At one time, web portals were nothing but a black hole for resumes, but now they are the only gate some companies will let you through. Keep in mind that they're looking for the right employee, not the right GPA. Key words are very important here. Your best chance for getting noticed through a web posting is to litter your resume with every key word applicable to your interest.

Amber was graduating in aerospace engineering with good but non-glamorous experience and a 3.2 GPA. She assumed her GPA meant the shotgun approach over the internet wouldn't do any good, so she focused on her University Career Services and networking. She had a connection into an aerospace facility in her hometown, and they required all applicants to go through the internet, so she posted her resume as recommended. The next week, totally out of the blue, and totally unrelated to her networking, she received a call from that company's rocket integration division two states away in Huntington Beach, CA. Her experience was exactly what they wanted. She didn't even know her experience had anything to do with launching satellites on rockets. She took the job, and thoroughly enjoyed it.

Every place likes to approach recruiting differently. Some places want every applicant to send a set of work samples (portfolio) before bringing them in for an interview. Part of this is to help match them to the right position, but also to see that the candidate is more than passively interested in the position. They're looking for somebody who wants the position enough that they'll follow up, and really make an investment in getting the job.

A cover letter can set you apart, and gives you another opportunity to illustrate your interests and understanding in reference to the company. Each cover letter you send will be even more customized than your resume. It takes more time, but if you really want the position, it gives you one more edge.

When you start applying, it can be helpful to keep a record of the companies you apply to, when you apply, and which resume you send. (Remember, you'll send a different resume to John Deere than you send to Motorola.) This record can help you keep track of when to follow up.

However you approach it, get your resume out there, and the interviews will come.

Telephone Interview

Telephone interviews are fairly common, so let's take a look at them.

It's hard to tell what a company's intentions will be with a phone interview. They may have looked at your resume and already decided they want to bring you in for an interview. They're just calling to check that you're not an absolute dolt before they fly you in. Maybe they're considering you and five other candidates. A phone interview is so easy, they call each of you. You never know. Frequently, interviews for internships will be conducted over the phone, and they'll decide based on that and your resume.

In some cases, a phone interview could just be a formality. In others, it could be very influential. Your best bet in all situations is to treat it as important as it might be. Here are some things to consider.

Don't talk too fast. All you have is speech to communicate. You don't have any pictures, hand gestures, or eye contact. The person on the other end of the line needs to hear your words and understand them. Make a conscious effort to keep the pace that you speak at a good level for communication.

Articulate. Don't slur your words. If you're not talking too fast, this usually isn't a problem. If you have an accent from your native language, from the South, Northeast, or wherever, you need to be even more conscientious about speaking clearly, so that the interviewer on the other end can understand you.

Be careful with humor and sarcasm. You don't have to be a stiff, but jokes don't always go well over the phone.

The interviewer will probably have a list of questions for the interview. This can make the interview seem rigid and formal, but that's just the nature. It is a good idea for you to use these questions to illustrate your experience and ability, just like you will in any other interview.

At the start of the conversation, your interviewer will probably tell you how much time he estimates the interview will last. Make a mental note of when this time runs out. He'll be flexible with it, but this can give you an indication about how much depth he's expecting with your conversation.

Often the interviewer will be an HR representative who is not closely familiar with the position. He'll make notes on an evaluation form, and the manager will make decisions based on these notes. In this case, it is important that you use experiences from your resume to answer the questions. Don't assume that the interviewer understands what you have done. He'll be looking at your resume, so reference items on there, and fill in more details. Examples are more effective than adjectives.

Don't ask about pay and benefits at this point. The rule is the first person who brings up compensation loses the negotiation.

University Interview

Several companies schedule interviews before showing up at career fairs. They always schedule these through the university's career services department. If you want a shot at one of these interviews, you must be on the ball having your resume in their electronic system, and applying to the appropriate companies.

One of these pre-scheduled, hour-long interviews could be a preliminary interview, or it could be the one where hiring decisions are made. Some companies allot a set number of positions to be hired from each specified school. They want three from Michigan, three from Stanford, two from BYU, two from UCLA, and so on. Since they know how many candidates they're pulling from each university, they feel this interview is sufficient.

Prepare for this interview just like you would an interview at the company. Dress appropriately. Be on time. Bring a couple copies of your resume. Bring your portfolio in a nice thin leather cover folder. And finally, relax. It's always an awkward feeling to be on campus in a suit in the middle of the day. Forget what you're wearing. Let your mind drift over your engineering experience. Don't think about what you'll say, just think about what you've experienced. Remember your interviewer is a person too and has probably spent an entire day talking technical. A few minutes of light conversation during the introduction will help relax both of you. Let your mind wander from one aspect of your projects to related aspects, and you'll be just fine.

If it's a cattle call style interview session for everybody willing to wait in line, go ahead and give it a shot. If a company is using that style of interviewing, they are doing so because they have found it to work for them. They won't make hiring decisions from that interview. They'll decide who to fly out for interviews.

You have a few minutes to make an impact. They'll interview scores of candidates, so get right to the point and give them something to remember your skill. You'll give them your resume to write notes on. Expand on the items on your resume. They'll probably ask about the experiences on your resume, one at a time. Fill them in with the same pattern you wrote the resume. Give them the Big Picture, then bring out the technical challenges. They'll be listening and scanning at the same time. It makes it easier for them to understand if you reference the first bullet, second bullet, and so on.

Don't finish the interview by saying, "I'm really interested, so go ahead and put five stars on the back, or however you mark the best candidates." When you make a comment like that, whatever good impression you may have made just got squashed. Seriously, think before you speak. The interviewers are already looking for the best candidates. They don't need to be reminded.

Company Interview

Most companies will bring you in for an interview before making a hiring decision. They may have already held a preliminary interview, or maybe not. This could be an all-day, eight or more hours series of interviews.

Show up on time. Being a few minutes early certainly won't hurt. If you're in a new city and you flew in the night before, it's not a bad idea to drive by the company that evening. Some of these big name companies have locations that are obscure and difficult to find. It's better to be lost the evening before than the day of your interview. Before you even fly out, following the directions on Google Maps can answer unexpected questions.

Get a good night's sleep the night before.

When you arrive, HR will probably see you first. They'll have some paperwork for you to fill out and sign. They may have mailed this to you earlier with your travel plans. If they did, having it filled out previously will keep your mind free for more important things during the day. If you have questions about things in the paperwork, then just leave it blank until you get there and can ask them to clarify.

They will have tried to schedule interviewers, and hand you a schedule before they send you on your way with your first appointment. In some cases, they may have had trouble lining up a full schedule. If this is the case, don't take this personally. It's not because they're less interested in you. Companies are busy, and chance can land you on the wrong day. Don't worry about it. Let them take care of that. Offer to follow up with people that missed speaking to you at a more convenient time.

Throughout the day, they may have each interviewer take you to the next, or they may have their HR staff pick you up and drop you off between each appointment.

Be conscious of the time and how long you're spending with each interviewer. You don't need to be looking at your watch, but if you happen to notice it's five past the hour, five minutes after your next appointment was scheduled, they'll appreciate you pointing that out. It's unlikely that they'll be holding a rigid schedule, so flow with whatever comes together.

They'll probably take you out to lunch. Custom dictates that they will pay, so don't think twice when they pick up the check (but have enough cash in your pocket...just in case). Order whatever looks clean and isn't red. You don't want your 3:00 interviewer to see on your shirt that you ordered Ottovio's broiled chicken in marinara sauce. Choose something that will allow you to eat and talk. Don't order alcohol under any circumstances, even if offered. Remember your manners and thank them for lunch.

You're at lunch, so you don't have to talk about engineering. You'll probably head out to eat with a few other engineers. This is their break from engineering for the day. Ask them how the skiing is in this area, or talk about Monday night football. Whatever. Leave your suit coat in the car if you want.

The more people you can talk with, the better. You're getting a feel for the company, and they're finding where you will fit in best. They're not looking to find things wrong with you. They assume that you really are the outstanding individual that you know you are. They're looking for where you will work best.

We'll roll through what the actual interviews may be like in the next section where we look at some real scenarios. For the most part, you're going to spend the time talking about practical engineering.

Back in the old days, apparently they would actually test an engineer's ability for academic performance during an interview. They'd sit them at a desk with a few problems, and come back in an hour to see what they accomplished. If any companies do that anymore, they're few and far between. Most interviews are more practical, focusing on the fit of the candidate with the team and the project. They expect that your university has adequately taught you the academics of engineering. Even so, some still like to push a candidate's technical ability.

One group of engineers likes running a good grill session. Three of them will all team up to interrogate one candidate. They have a good time with it, as do the candidates. They won't bring in candidates unless they're confident that they can at least hold their ground. If you get pulled in to get grilled, they're probably already confident that you'll do well.

No two sessions are the same. Since all engineering is interrelated, they can start anywhere, and by the time they're done, they've pushed to the edge of your knowledge in structural analysis, materials science, machine design, numerical methods, thermal analysis, and anything else they can find. They don't just talk about theory. They have a white board and plenty of paper. They ask you to write the equations, balance the equations, and even pull some numbers off the top of your head where your experiences says you should be familiar with them. They'll draw some lines on a chart, and expect that you recognize what those lines are most likely to represent, and what it means with respect to the analysis you've been doing on the board.

If you land in one of these sessions, it can be the highlight of your whole day. If you brought your portfolio with you, including some technical work you already did, they'll dig into it, and since you know what you did with that analysis, it's a good opportunity to show off. But beware: they will dig deep enough that you'll be left saying, "Hmmm...not sure on that one." At that point, they'll throw you a curve from a totally different angle, and you have to switch gears on the spot.

You may find yourself in an interview where the manager will describe a design problem, tell you to brainstorm solutions, and leave you in his office to scribble feverishly. Fifteen minutes later he'll walk back in, look over your paper, crumple it up, and tell you to come up with more ideas. He's probably looking to gauge your creativity and problem solving, seeing what breadth of ideas you can bring to the team.

And there's always the possibility that you'll end up in an anomalous situation that you just can't prepare for. Sara finished an interview that went very well. Her experience precisely related, she aced the brain teasers thrown at her, their personalities matched...and then came the question.

"Well, you seem like you'd be a great fit for the position. I'm sure you'd do very well, but there's just one more thing. Are you *in like with programming*, or are you *in love with programming*?"

"Umm, what does it mean to be *in love with programming*?"

"Do you dream about programming? Do you have a C++ manual by your bed so when you wake up in the middle of the night, you can check if something will work? Do you think about programming while you eat? Do you think about programming while you drive to work?"

Sometimes, you just have to accept that you might not be able to deliver what the interviewer is asking of you.

People have all kinds of interview war stories. The best recovery is composure. Stop for a moment, take a breath, and let your mind relax. If you think you just said something stupid, acknowledge that you just said something stupid, and let it be. Pick your mind up again and move on.

Maybe you walk out of an interview with one individual, and you feel like you choked. It probably didn't go as bad as you fear, but remember that the next person you'll speak with has no idea what just happened. You're starting with a clean slate again, so let go of whatever is past. Stay positive and keep your interest in the work the company does.

You must apply if you want to get an interview. Register with your University Career Services on the first week of the semester. Submit your resume online. You may find yourself in several different interview scenarios. Prepare for each accordingly. If you slip up at some point in the interview, regain your composure, and carry on.

Chapter 5

How to Speak Tech–*Interviewing Skills*

Interviewing is a skill that can be learned, just like composing your resume. It's built primarily upon communication, technical communication in particular. If you learn to interview well, the same skills will carry over to other areas of communication. We're going to walk through some of these skills, looking at examples of them in action.

The primary purpose of an interview from the perspective of you, the candidate, is to show the company that you will be a constructive member of the team. From the perspective of the interviewer, the primary purpose is to find the best place for the candidate on the team (even if the best place happens to be off the team entirely...but we'll stay positive here).

Now we'll start into the tactics of interviewing, first taking a look at attitude.

Attitude

Everybody who gives advice about interviewing will always say, "Be positive, and put on a smile."

Most people forget that advice, and do well in an interview. Then there are those people who remember that advice...and it's sickening. The only thing worse than a negative and complaining candidate is one who is unbearably excited, forcefully happy, and walking around with a ridiculous smile plastered on his face. It's unnatural. I've never interviewed anybody who came with a negative attitude. I have interviewed a few people with this cheese-ball attitude. Don't do it.

Having been cautioned against overbearing happiness, it is true that a positive attitude and a smile is something we look for. So how do you portray this without being a goon?

We want people who are easy to work with and who are excited about the project. We want individuals with a positive can-do attitude. This is different from a permanently beaming smile. This is excitement and positive confidence associated with engineering.

You show this positive attitude by being interested. Yes, this principle of *being interested* is getting monotonous, but it really is the core of what makes a good engineer. Get used to it, because we're going to keep hammering it in. We'll get back to this in a moment, but there is another aspect of attitude we want to cover.

Avoiding negativity is important. Stay positive by being interested in constructive topics. Don't be interested in paperwork slowing down the design process. Don't be interested in the inability of a group to make decisions. If your interviewer comments on any of this, acknowledge what they say, and let it lie. If the flow of conversation dictates that it's your turn to speak, ask about the technical nature of the decisions these groups are addressing. "What kind of analysis is important to knowing if the optical sensor got contaminated? How would you go about fixing the servo motors if they actually find out that the bearings are bad?"

Avoid questions that dig further into the cynicism. "Why can't they come to a decision? Isn't there usually a clear answer to what they should do?"

If you're asking questions about challenges with the people or the paperwork, you're probably leading to negative conversation. If you're asking questions about challenges with the hardware and engineering, you're probably leading to positive conversation.

If an interviewer is cynical or negative about his job, about the company, or about the project, that should raise a yellow flag in your mind.

While talking with engineers and managers during your interviews, you will have several opportunities to pursue lines of interest, or leave them to rest. Keep your interest constructive, and your interview will go well.

Take an Interest in What They're Doing

Your resume got you into the company for an interview. They've investigated what you've done; now they want to investigate what you will do. They are looking for somebody who fits onto their team, and who fits into their projects. They've seen that you fit into other places; now they want to see you fitting into their place. When a company brings you onto their turf, they're interested in how you will perform on their playing field.

The most effective conversation you can start with is about what they're working on. Their current projects are more interesting to them than your old projects. Compounding that, their current projects are what they're hiring you for, not your old projects.

Take an interest in what the individuals you are interviewing with do. Ask them about their responsibilities and about their current projects. Ask them how different aspects of their projects work, technically. Don't hound them with questions. If they're not interested in talking about some aspect of a project, don't push them on it. Just let it flow like a regular conversation.

Your interest needs to be genuine. You don't have to be interested in paperwork, because you're probably not interested in that. If you're not interested in solid modeling and using CAD programs, then don't be interested in that. If you *are* interested in numerical analysis or structures, then *be interested* when you get an opportunity to talk about it. If you *are* interested in motion control, then ask questions related to the motion control challenges they work with in their projects.

You will choose what you're interested in (or you'll spontaneously become interested when you see something you hadn't previously considered), but you don't need to be negative about things that don't pique your interest. If you are not interested in something, simply don't dig to understand it any deeper.

If the interviewers are pushing something you're not interested in, flow with the conversation, and start understanding why this is important to them, to the project, or to the company. The more you understand, the better you'll be positioned to perform effectively. They may pick up your disinterest in the activity they're talking about, but they'll recognize your interest in being an effective team player.

Engineers draw pictures on boards and tape diagrams onto walls. They'll try to clean up their office before a candidate comes to interview, but there will still be drawings and charts sitting around. If you see something interesting, ask about it. Don't snoop around, but if it looks like something they're working on, they'll probably be happy to talk about it.

Inquiring about their work will serve three purposes. First, you'll get an idea of the Little Picture work that goes on at the place. If you're interviewing with a fairly young engineer, it will give you an idea of what kind of responsibility you can expect.

The second reason you take an interest in what they do is to start gun slinging that technical mind you brought with you. You can start building an increasingly detailed understanding of some aspect of the full project. Why are the data represented on that chart important? Are they from a mathematical model or actual test results? What kind of variables did the mathematical model account for? What kind of equipment was used to read in the test data? How does that line up with the expected performance? What factors caused any anomalies in the data? All engineering is interrelated, so you can take one starting point and go a long distance with it.

Remember, don't hound them with your questions. As they start talking about what they're working on, they will naturally focus in on the things they're most interested in. Follow their interest. Ask them to clarify or explain concepts that you don't understand. They'll be happy to explain the brains behind their work. As they communicate their understanding to you, they'll see that you are a good communicator, a quick learner, and an interested candidate. They'll feel smart, and in turn, they'll feel that you are smart. Most important, they'll begin seeing where your mind would be most effective within their organization.

The third reason you take an interest in their work is to lead to opportunities to show off the work you are familiar with. Follow your interest in their work as far as it naturally goes. Along the way, you are bound to run across three or four points that relate to your work. When it is time to talk about your experience, you will take one of these points, and lead a tangent path connecting to an aspect of what you have been involved with.

Relate Your Experience to What They're Doing

My favorite interviews are those with candidates who find and take opportunities to relate their technical experience to what we're doing. It's an easy skill to learn. Usually, a candidate who has learned how to do this will take an opportunity within the first 5 to 10 minutes.

When we're talking about the satellites we work on, the testing we put them through, or the equipment we use for the testing, they'll pick up on some detailed aspect of how the systems work and relate that to an experience from their resume. That makes it easy for interviewers, because we can then follow them into their technical background. We become interested in what you've done because it relates to what we're doing right now.

As an interviewer, I like to learn from the candidates we bring through. You're just graduating, but you have worked on things that your interviewer may not even been aware of, let alone understand. You have developed an expertise in some obscure technology, or learned the technical challenges in some aspect of an industry.

It's exciting to talk with somebody who has been working in a totally inconspicuous field, developing a technology that nobody would think about unless he were in the thick of that miniscule industry. Most Senior Design Projects are of this nature, as are most projects students work on with smaller companies for the summer. They're seldom involved with high glamour projects like aircraft, but the engineering behind the solutions they create is always interesting.

One candidate in particular, in 15 minutes, taught me more about the functional design considerations of more types of valves than I knew existed. This candidate had spent a summer working on a diamond tipped high cycle impact piston for smashing asphalt. I don't recall what his actual activities were, but he understood the principles of fluid flow, pressure transfer, challenges with high speeds, high pressure, impact, and high cycles. He understood materials, and design considerations in choosing their proper application. He understood lubrication challenges within an inherently dirty system. He even understood considerations for long-term maintenance by field operators.

Every practical aspect of engineering associated with this machine was loaded into this guy's mind. He knew his project. It was clear from another numerical analysis project that he was thoroughly capable in the more theoretical aspect of engineering as well. Nothing in his background said "Aerospace" or "Satellites," but I would be entirely confident to bring him onto any team working on any project. He showed that he can clearly see all the Little Pictures, and how they interact within the Big Picture.

It wasn't a fluke that we ended up talking about his project. We do satellites, which has nothing to do with smashing asphalt. Using the skill just identified, he brought up the topic.

While talking about the engineering in the test chamber, I routinely mentioned the valves, and he took the opportunity to tie it into his project.

> "What kind of valves do you use?"
> "Some gate valves, solenoid valves, proportional valves for the LN2. Just typical valves. Why?"
> "We looked through a lot of valves for our asphalt trenching project. We had to find some valves that were fast enough, and could switch under high pressure."
> "Oh yeah? What were they doing?"
> "Well, we had to fire these pistons at seven hundred pounds, at four cycles per second. When I started, they were developing a system using...."

And the conversation flowed from there. I followed his interest to see how deep and wide his understanding went, and he kept tying in new design considerations. He didn't have pictures with him, but we drew several sketches in the process of talking about how everything actually worked.

You can gauge somebody's ability for technical understanding by how much he's able to teach you from what he worked on. It doesn't matter what he worked on. It just matters that he understands it, and that he is fluent in communicating about it.

When you walk in for an interview, be aware that through your projects, you have developed a relative expertise in a few areas that your interviewers know very little about. Be prepared to teach them the engineering involved in the projects you worked on. Not just *your engineering*, but *all the engineering* in the project.

You don't need to have actually performed all the engineering you talk about. Understand it, and be ready to talk about it. At one point during our conversation about this asphalt machine, he stopped and said: "I didn't actually work on this part." My response: "That's fine. I want to see what you understand about it."

If you are afraid that you may be inflating your experience, just use the line "I didn't actually work on this part, but, the system requires. . . ." You're being honest, and you're demonstrating your understanding.

Controlling the Interview (Your Objective in the Interview)

We've all heard the advice to "take control of the interview." This is what we just demonstrated here. You actually *share* control of the interview. However it flows, you need to keep your objectives in mind, and create opportunities to illustrate your technical experience.

Suzy commented about an interview that she felt had too much emphasis on the personality side.

"We got along great, and had a good time talking. But when I walked out of her office, I realized that we hadn't talked about anything technical at all. We could have had the same conversation even if I wasn't interviewing."

Her concern was that when this interviewer went to fill out the evaluation form, she wouldn't have anything to gauge her technical ability or interest. For the particular job for which she was interviewing, Suzy knew her technical aptitude and ability for technical communication was highly important, and she needed her interviewers to know she was up to the challenge.

This can happen to any candidates who rely on interviewers to ask questions about their past experiences, or about how they would approach solving a particular problem. Usually interviewers will bring up something technical, but it's not always the most effective piece of experience that needs to be shared.

Interviewers will eventually ask you about your experience in a generic way, if you don't bring it up in a relative way. They'll eventually ask how you would go about solving some hypothetical problem related to their project, if you aren't already investigating how they're solving their current project challenges.

It will be far easier for interviewers to see you as part of their team if you are naturally inquiring into their project, and jumping right into evaluating the technicalities that are important to them. If you keep your objectives in mind, you will be able to help your interviewers achieve each of their objectives, and do so in the most effective manner for both you and them. You will be able to help guide them to seeing your technical ability in reference to their particular projects.

Additionally, it will give a much more accurate picture of just how easy it will be for you to be interested in this job. You will be showing up to work at this place every Monday through Friday for at least the next two years. You want to know how easy it is for you to be interested in the work.

Keep in mind that some interviewers may have a set routine they like to follow. They may have traditional questions about your background. They may have technical challenges they use on every candidate, and they like to compare how different candidates approach them. If this is their style, roll with it. If they're more interested in asking a particular question, then float with their interest.

Each interview is a joint discussion between two engineers, and as a result it will reflect the joint interest and personality of both engineers. All you're doing is following each other's direction of interest in a normal (technically focused) conversation.

You will probably not cover all of your resume experiences in each interview. You may even talk about some experiences that aren't on your resume. You will probably not deal with each experience to the same depth with different interviewers. It is more important that an interviewer sees the quality of your understanding, rather than seeing the quantity of your experience.

How to Talk About Your Experience

How you talk about your experience can be more important than the experience itself. You have already seen on your resume how much difference there is between an effectively presented experience and a poorly presented experience. It's just like dinner at a nice restaurant. The food really isn't five times better, but the presentation certainly is.

Earlier, we looked at Bryan Hyatt's experience testing missile systems, and saw the difference between how he initially spoke about it and how he learned to speak about it. The experience hadn't changed, but he went from perceiving it as a flop, to understanding it was a very valuable and technically involved project. It was reflected in both his resume and his interview.

Illustrating challenges with understanding Some tactics for talking about your project can significantly influence the way challenges illustrate your ability. You want to speak about challenges, without speaking negatively about the project. At the same time, you want to speak about challenges in a way that shows you understand what you were faced with. Compare these two illustrations of the same technical challenge on a project.

"We had a hard time getting our data acquisition equipment to work, but eventually we sorted it out and got our data."

This shows frustration and discontent with everyday problems, without showing any understanding. Instead, talk about challenges by explaining the problems and how you overcame them.

"We didn't have the updated drivers for our data acquisition equipment, so when we figured that out and installed them, we had no problem collecting our data."

Each of these explanations takes the same number of words, but the latter shows a whole level of understanding and confidence that the first totally bypasses.

Recognize that the "simple and stupid problems" you run into with your student projects are very similar to the simple and stupid problems engineers deal with professionally. We want engineers who are competent at working through these simple and stupid problems, without getting frustrated, so they can focus on the "complex and challenging problems." Seeing that you understand these mundane problems gives confidence that you won't be slowed down by them.

Carrying the conversation How you talk about your experience isn't just in the words you choose, but in how you carry the conversation. We'll look at this in an interview with another candidate.

Eric came in to interview for a summer internship. He has an impressive mind, but hadn't yet learned to effectively lead an interview to bring out his technical understanding. He made the classic blunder so commonly stumbled into by many undergraduates; he downplayed his experience.

He had a well written resume. Comprehensive, technically articulate, concise, everything we look for. I asked him about his experience with optimization using a cutting edge development of Genetic Algorithms called Evolutionary Epochs. Optimization is one of my favorite fields, and the moment I saw this new thing called Evolutionary Epochs, I was looking forward to learning about it.

Here we are, we've been talking about his other experiences as the conversation flowed, and I finally got a chance to ask about this new aspect of optimization.

> "So, tell me about Genetic Algorithms. What were you guys doing with this?" Visualize the restrained anticipation.

"The optimization? Actually, that wasn't very involved. I didn't really do much with it. I was pretty much just a number cruncher. It was this graduate student's project, he was doing all the research on it, and we just helped."

Visualize the traumatic disappointment.
"Ok. . . ."
Visualize searching for where to take the conversation.
"So. . . What are these Evolutionary Epochs? How do they tie into Genetic Algorithms?"

"Evolutionary Epochs are associated with each of the sets of sequential generations as the algorithm progresses. You run so many generations, and systematically select to retain certain patterns that repeatedly fit subcriteria of the overall model. But that's what the graduate student was working on. That's not what they had us doing."

"That's fine if you were just crunching numbers. I don't care. How would you recognize these patterns within the full model with each generation?"

"We would take the new solution, evaluate it for. . . ."

It turns out Eric had a functional understanding of the principles this graduate student was working with, and could speak about it very articulately. He wasn't sure in some areas, and acknowledged when he didn't know, but offered his guess as to how the algorithm worked, and that was interesting as well.

This conversation thoroughly illustrated that he understands numeric analysis, can pick up on complex logical patterns, and will understand the Big Picture of what he is working on. The terrible thing is that the conversation almost didn't even happen. He almost killed it before it got out of the gate, and then tried to shoot it down a second time. As the interviewer, if I hadn't been interested in optimization and known what questions to ask, I would have got the impression that he half-fudged his resume. Instead of being a strong positive indicator, it would have become a yellow flag.

We can look at a critical review of how a candidate should approach this kind of situation in an email I sent Eric a few days after the interview. The second part of the email will lead into our next topic of discussion.

Like the outstanding candidate that he is, Eric sent the appropriate follow-up email the day after the interview. I replied, asking if he would mind his resume being used as a workshop example (it really was well done). He said that was fine, and asked for some suggestions on things he could improve from the interview. The email is fairly blunt, but he took it with the constructive attitude intended.

> Eric -
>
> Thanks for permission to use your resume. I appreciate it.
>
> Feedback... I'll state up front that I was impressed with you, wrote a good evaluation, and hope to see you here this summer. You can feel good about the impression you made. That being said, I'll be blunt about what may improve. But again, this is just one opinion. Evaluate it and make your own judgments.

Resume: Group your Lab and Research experiences under Engineering, and your other stuff under another heading. Other than that, your resume is great. Actually, if you make those revisions, fire it back and I'll use that one instead.

Interview:

Don't downplay your experience: You tried to convince me that your research on the Genetic Algorithms wasn't very significant. Sure, it wasn't super complex PhD level development. We don't care. You're only an intern. What we want to see is that you understand what you were working with. Instead of ending the topic by saying "it wasn't that involved", use that as a launch pad to talk about Genetic Algorithms. Tell me how the presence of Evolutionary Epochs are important. Ask about my level of familiarity with Genetic Algorithms, and about Optimization in general. Find what my exposure has been, and then bring me up to speed so that I can understand what you "worked on." Not necessarily "what you did" (although I do want to know that as well), but what it is you were "working on." Tell me about the project and its goals. Of course you don't want to inflate your experience, but you do want to illustrate, as comprehensively as possible, your understanding. We want to see that you have the ability to understand the larger project that you are engineering. Then we want to see that you have the dedication to pull through on your aspect of the project that you are individually engineering. Again, don't inflate anything regarding your role in the project. You don't need to. Talk about the project, how it works, why it's important, what the challenges are, etc. Tell us what you individually did, what was interesting and challenging about it, etc... then tie that into the larger picture. Don't tell us that your experience wasn't very valuable. Show us the value of the work you were involved with.

Bring a few samples (papers) to use as illustrations: I just finished interviewing a candidate from Harvey Mudd, and he blew me away with his detailed understanding of the interactions between components on circuit boards. He pulled a few diagrams and papers out of his folder as was appropriate, and that really helped my understanding of what he had worked on, and gave him an excellent forum to illustrate his understanding. Socially, he is soft spoken (more so than you even), but he is fluent speaking technically. It's not that his project was any better than other EE's projects. The difference is that he knew how to talk about it, and how to illustrate it. His GPA is only a 3.2, but I have no concerns about his technical ability. Find ways to bring the conversation around to technicalities, where you get to talk about the physics, science, and engineering involved with what you worked on. We just want to see that you can wrap your mind around technical problems, and communicate about those challenges. Pictures, drawings, and charts are useful for talking about technical things.

So that's that... I guess. I hope it's helpful. I'll be happy to answer any other questions you may have. Before this goes any longer... good luck, and thanks again for permission with your resume,

~Bryan

You want to lead the conversation to subjects that allow you to communicate your engineering understanding. When interviewers bring up one of your experiences, instead of commenting on how you feel about the project, go right into the engineering. First, paint the Big Picture so that they can see the overall setting that you're working in. Describe the project, what its purpose is, the primary goals, and what the key technical challenges involved are. From there, move into the challenges you personally dealt with. They'll ask questions as you go, so take those questions and fill them in with all the interesting stuff you have packed into your brain. Be sure to identify any acronyms and other jargon you use, so that your interviewers can easily follow.

Everything relates to something else, so travel from design factor to design factor, illustrating your project, and tying it into their work whenever it's appropriate. You want them to see how you will fit into their team, so sell as you tell.

Engineering Portfolio

Your portfolio is a communication tool to use in illustrating your interest and experience. As briefly noted before, your portfolio will be a powerful tool in an interview, and can also be a powerful tool for getting an interview.

When some engineers hear the word portfolio, they imagine an artist pulling out a square meter folder with watercolor paintings, oil paintings, and artsy looking photographs. This isn't too far from what your engineering portfolio will be like. It will be a collection of visual representations of your work.

What actually constitutes an engineering portfolio is about as diverse as engineers themselves. It could be nothing more than a few pictures, drawings, and calculations. On the exhaustive side, it can include project descriptions, reports, samples, and everything else under the sun. You need to create and bring the portfolio that is appropriate to you and to the interview.

Posting your portfolio on a web site can be worthwhile. Posting it on a web site opens the scope of your audience, so it needs to be put together such that it doesn't narrow your interests. This isn't difficult to do. Just write positively about your experiences.

To compile a portfolio, all you need are one or two pictures of your projects (more if they're interesting). These can be photographs, schematics, CAD drawings or models, etc. A picture is worth a thousand words. You're looking for something to instantly communicate those thousand words of explanation.

Including a report from one of your class projects can be effective. Many managers are interested in a candidate's ability to write. Believe it or not, your professor isn't lying when he says writing will be important (arguably, writing ability will be more important than technical ability). Furthermore, many managers argue that it takes more understanding to write about a project than to talk about it. They'll be impressed if you can clearly communicate in writing. It doesn't matter what you're communicating about, it only matters that they can follow your train of thought. If the project report was a group effort, mark the content you were responsible for. With a group project, you don't misrepresent your ability, and you want an interviewer to easily find a representation of your work. If your professor thought the report was good, your manager probably will think likewise.

It is usually best to compile this portfolio electronically. Scan in your pictures, keep screen shots, etc. PowerPoint or Word are good for this (the Microsoft or PDF formats make it accessible by anybody).

You want to have everything printed on paper for your interview; it's easy to keep organized, you can draw on them, and leave them with your interviewers. If you are going to email your portfolio to anybody, a simple PowerPoint presentation with the pictures and a few descriptions is very effective.

If you want to create a more exhaustive portfolio, you're in luck because you've already generated most of the writing in your Experience Document that you created for composing your resume. All you need to do is sort in a few pictures with these descriptions, and you have a comprehensive portfolio.

Using your portfolio In your portfolio, what you have created is a communication tool. You will use it in person during your interview, and you may also use it as an addition to your resume, to help you land an interview.

Having a picture makes it easier for you to communicate your train of thought, and for your interviewers to follow your train of thought. It makes it easier to point at one part of the project, and then point to another, providing seamless communication about which technical parts you are referring to, without encumbering the conversation with identifying each part each time. It also makes it easier for your interviewers to understand the full project you were involved with. The more clearly they understand your project, the more they will judge that you are an effective communicator.

On the technical side, pictures, drawings, and equations allow us to see your technical ability. With pictures and drawings, we get to see *real engineering*. As simple as you may fear it is, it really makes your project come alive, and the technicalities come alive.

A picture also allows many peripheral skills to become apparent, without even speaking of them. If you pull out a drawing, we don't even need to mention tolerance, and the tolerances on your drawing will show you are familiar with that. If you have some simple force balance equations sketched below your project picture, your interviewers can see that you have done some engineering analysis. They may ask about the equations, which gives you another opportunity to let your engineering mind shine.

You will probably not use every aspect of your portfolio with every interviewer. Let the interview go in the direction that it naturally flows. You simply have these tools to help communicate when the interview flows in those areas.

If you have accomplished significant things that are not directly engineering related, having part of this in your portfolio can really turn it into a strength. Tyson Mao came through to interview with us, and he had typical engineering experience that comes with a degree, but he had spent an unusual amount of time re-instigating National

Rubik's Cube Competitions. The Rubik's Cube fad that had flourished in the 1980s had died off, and Tyson has been bringing it back.

That shows a measure of self-motivation, and in the engineering direction, algorithm development. But what really set him apart was the Rubik's Cube Competition Rules he brought with his portfolio. He had personally drafted this 60 page document defining all the competitions, how they are to be carried out, all the rules and provisions for judging what is legal and illegal, etc. It was tedious, exacting, and meticulously precise...it was great! It looked exactly like the assembly, testing, and operating procedures we create and use for everything carried out on our satellites. Here was a candidate who, on his own, had developed an appreciation for the rigorous detail necessary for aerospace paperwork. With this kind of aptitude, Tyson would make an excellent systems engineer, and he would enjoy doing it.

It is worth pointing out that Tyson's interest in Rubik's Cube competitions wasn't the impressive factor; it was his skill demonstrated in following through. We didn't speak about the marvels of the Rubik's Cube. Rather, we spoke about what it takes to organize people and their activities on the level that he has done. It was clear that Tyson's first interest was aerospace, and he had demonstrated skills that are valuable to the aerospace industry.

You will probably not leave your portfolio with people to review later (especially if they're senior managers with many demands on their time). Your portfolio will be a communication tool to use in person, and only be left if the interviewer asks for a copy.

An additional benefit of putting your portfolio together before the interview is that it serves to remind you of the technicalities of your projects. It reminds you that, yes indeed, you have been training to become an engineer during the last four years. This refreshed familiarity with your projects helps you to communicate more proficiently during your interview.

During an interview, be interested in the projects the company and the engineers are working on. Relate your experience to projects they are engaged in. Illustrate engineering challenges with understanding of how you worked through them. Don't downplay your experience. Talk about what you understand of the Big Picture, in addition to what you did in the Little Picture. When appropriate, use your portfolio as a communication tool.

CHAPTER 6

Other Interviewing Considerations

Networking

A highly overlooked tool in the engineering field is networking. Networking among engineers isn't the business schmoozing style of activity that some students associate with the term "networking." Networking among engineers is very simple. It consists of talking to other engineers. It isn't asking for jobs and interviews. It's just talking with other engineers, about engineering. Another business word for this is an "informational interview," but even that makes it sound more formal than it is. The person doesn't even have to be an engineer. Networking is just talking with people about what they do in their job.

Let's take a look at networking from the perspective of the engineer out in industry.

Imagine you graduated six months ago, you're working for Ford in their suspension testing department, and you get a phone call from one of your friends back at school who worked on the Human Powered Vehicle with you. You worked on this project together, but you weren't exactly buds that hung out on the weekends. He got your number from a professor, and you're glad he called so that you can say hi and chat a bit about life and what's going on back at school.

Now this friend takes the conversation in the engineering direction.

"Hey, I was hoping to chat a bit about what they have you doing there in the sense of engineering. I'm going to start interviewing with different places and trying to land a job, so I was hoping you might be able to give me an idea of what to expect."

High pressure? Not at all. Just good friendly advice from your own experience.

When you're sitting in that cubicle, you're not going to feel pressured at all in talking about what you're working on, or what you like about your job. You know you don't have the power to get your friend a job, but you do have the power to share what you've experienced. You also have the power to talk about what the other new engineers you work with generally spend their time doing.

Your friend didn't ask you to get him a job. He just asked about the job you already have.

Your friend takes an interest in your job, and carries a normal conversation about it. You laugh about the quirks of a corporation, the occasional dull times, but he stays interested in the work that you're actually doing. He wants to know what you do so that he can know what kind of job he wants to focus on.

Then after chatting, your friend asks you if you would mind looking over his resume. You say, "Sure, email it on over." Your friend didn't ask you to pass it on to your manager, or on to HR, just to look it over. You're happy to do that, because that doesn't put you at risk for anything. While you're telling your friend to send his resume, you're already thinking about who to send it to within the company, but your friend hasn't said anything about that, so you're not feeling pressured.

Now your friend says, "I put a project on there from a class I took from Dr. Sorensen on Manufacturing Operations. It was an interesting class, and I enjoyed the project. It sounds like you're more involved in test than in manufacturing. Do you know anybody who is closer to the production line? Your job sounds interesting, but I'm also a little interested in what they have going on in manufacturing."

You know another new engineer and have met a couple managers who are over in manufacturing. You're comfortable interacting with them, so you offer to connect your friend to one of these people.

"Yeah, Sarah's working in the manufacturing area, and she seems to like it. I don't think she'd mind talking with you. Do you want her number? I'll let her know you're interested."

See how simple and non-offensive that is? It isn't demanding. It isn't asking for a job. It's just being interested in what people do. Call to gain exposure, and let the hiring take care of itself. That is networking.

Understand that you're not imposing on the engineers you call. Quite to the contrary, the engineers you call feel important. The most common statement coming from an engineer after talking with a graduating student is "A guy from my school just called and we chatted about my job for a while. It kind of makes you feel important, or professional, or something." I've never heard an engineer complain about a student calling to talk about his job. Never.

When you do begin networking, you will find that the exposure you gain to so many people's jobs is highly valuable. It will dramatically increase your field of interest, enhance your understanding of the industry you are investigating, as well as develop your fluency in speaking about your own experience. The networking is still part of the learning process, and part of the preparation process for the actual job interviews.

Again, when it comes to networking, you're not calling to get a job; you're calling to get exposure.

The Interviewer Perspective

Interviewers want you to look good. They want to deliver a positive evaluation of you. They want to be impressed. Keep this in mind. They're going to do everything they can to make your interview a positive experience, resulting in a good evaluation of you as the candidate, and planting a good impression of the company.

They aren't trying to weed out the underperformers. They selected you, among several hundred resumes, to bring in for an interview. They chose to bring you to the company for an interview, so they have already decided you are very likely a sharp engineer. They're looking to find the right match for you within the company.

Younger interviewers can tend to be a little bit nervous. They haven't interviewed many people, so they aren't sure what to expect. The questions running through their minds: What are we going to talk about for a whole hour? What do I ask him about? How do I tell if he's smart? What if we have nothing to talk about? What if he thinks I'm an idiot? Even worse, what if I think he's an idiot? What if I think he's good, but other people give him bad reviews? I must be pretty important if they want me to interview this candidate. That's cool. What if they think I don't do a good job interviewing? My cubicle is so small, how are we going to fit?

Engineers that have been around longer have been through this routine a few times, so they're likely to be more concerned about their responsibilities than getting any jitters about the interview. Just keep in mind that the interviewer already thinks you're a quality candidate, and he's going to do what he can to make it a good interview.

Managers will be inclined to take a more neutral perspective. They really want to see if you will fit into the team. In the back of their minds, they're thinking, "If I make a mistake with this person, I will have to live with it." Be honest and be yourself. The best way to ease their concerns is to let them feel that they are seeing the real you.

Follow Up

There are two aspects to following up. One is the initial thank you note, and the other is oriented toward the long term.

Always send a thank you note to all your interviewers. Get their card and send them a short email the next day. Thank them for their time, but remember this is another brief opportunity to express your interest in working for them. Reiterate your interest in their projects by mentioning a skill you discussed during the interview and applying it to their projects. For example, "I would enjoy the opportunity to enhance the CFD skills I gained designing micro-air vehicles by working on projects such as the engine intake system you currently have in test." Include something of substance, but make it concise without being limiting (like your resume Objective statement). Keep it brief, and attach your resume.

Make a phone call to the hiring manager a week later to follow up on the decision process, and see if he needs any more information. Stay positive and remember your manners.

Following up in the long term can be very beneficial. There are three things that need to match up for a job to work out: the right person, the right position, and the right time.

Following up is how you get by when the right position or right time doesn't exactly coincide with you being the right person.

Matt Keller started with a major aerospace company at the end of the summer. When Matt chose to work for them, he bypassed three other job offers. That makes it sound like he was in high demand, but the story goes quite differently.

Matt had been applying for jobs his whole senior year and nothing was coming through. A few places interviewed him, but none delivered any offers. As you can imagine, and may have to experience, graduating without a job isn't the greatest feeling in the world. He kept at the job search, applying online, checking with his University Career Services, networking, and following up with the places he had interviewed with. He was playing the game right, but a job wasn't pulling through.

Then toward the end of the summer, four job offers came through, all in the same week. Just like that, he went from zero demand, to high demand. All four offers resulted from following up after applying or after interviewing. He was the right person, but the right position and right time didn't match until several months later. Since he followed up, they knew he was still available and could quickly interview or send offers.

Answering for Your GPA

If interviewers comment that your GPA is lacking, don't worry about it. They chose to interview you, so follow up with the strength that they chose to consider you. "I've never done well on tests. It's the projects that I enjoy, and where I'm valuable." If you worked to support yourself through school, that can be worth pointing out. Answer their concern, and let it go.

If you are an academic superstar with a 3.9, and want to avoid the stereotype of the arrogant scholar, just brush it off and move on. "I just do my homework." Or just say "Thanks" and let it be. Either way, don't dwell on it. If they're impressed, they'll stay impressed. You want to focus on engineering.

Spend What Is Necessary

Put things into financial perspective.

You've spent how many thousands of dollars on your engineering degree? You'll be making how many thousands a year when you get hired? And you're not sure if it's worth a dollar a page to laser print some nice color pictures for your portfolio? It probably isn't outrageous to spend at least as much on a suit as you spent on the textbook for your Differential Equations class.

Employers will want to contact you. If you don't have a cell phone, seriously consider getting one. This allows you to get any messages they leave, and gives you the freedom to schedule a phone interview far more easily than running back to your apartment. Just remember to turn it off in class. And by all means, don't be the idiot who answers his cell phone during an interview. Turn it off, not just to vibrate.

What to Wear

There are a few bits of advice for both men and women. First, don't wear perfume or cologne. You never know what your future boss may be allergic to, or just dislike. It's best to just smell clean. If you wear glasses, be sure to give them the once over again just before you enter the office. Also, grooming counts, so make sure fingernails are clean and trimmed.

Guys: Spend some money on a good suit, shoes, shirt, and tie. Whatever you spend will easily pay off when you get hired. Don't go high end on the fashion. Stick with tried and true combinations like a white or pale blue shirt and classic tie. Remove any visible piercing. Don't stress about what you wear. Just make sure you're comfortable in it. If you despise suits, a shirt and tie will be acceptable.

For women, we'll call upon the advice of a good friend, who is far more versed in interviewing for women than any guy will ever be. If any of you guys are tempted to *not* read this because it doesn't apply to you...well, some things do apply, and it's worth reading just to appreciate how much more simple your life is.

Business suits always do the trick. Slacks may be preferable to a skirt, but you want to be cautious about sending any feminist overtones. You never know what your interviewer's idea of "conservative" may be.

Heels affect the way you walk, making you look more confident. However, confidence comes from inside, not from your shoes, so if you're not comfortable in them, don't wear them. You'll be on and off your feet all day, so be sure your feet are comfortable. Closed toes are always a safe bet. Open toes are less professional, and really bad if there is a chance you'll be around machinery.

Women should not wear more than seven pieces of jewelry (each earring being counted as one). Dangling jewelry can be distracting.

Wear your hair conservatively and out of your face.

Makeup should be simple. Stick to muted tones and neutrals. This is not the time to make a fashion statement.

A clear coat of polish or a French manicure is appropriate for fingernails. Save the colorful nail polish for after you've been hired. If you decide to wear open toe shoes, go for a toenail polish color that complements your outfit, but isn't too eye catching. Skip the decals and painted flowers on interview week, too.

Both purse and shoes always look better when they match each other and the suit. These are also items where it is okay to express a *little* bit of your personality. Leather is always the most professional choice. Aside from your purse, all you should have in your hands is a simple folio.

And here are two bits at the end from the guy's perspective of all these rules for women.

Whatever you wear, remember that you're still around engineers, the same kind you've been working with through college. If your outfit isn't just right, the men won't notice, and the women will be more interested in your interaction than your fashion. Just make sure your outfit isn't distracting to you, or to them. The most important part of your outfit is your smile…so wear one proudly! It will distract the interviewer from whatever you might have overlooked.

Choosing the Right Job

Several seasoned engineers have offered the same line-up of priorities to consider when choosing the right job:

1) your boss
2) your project
3) your salary

Your manager is the most important component because he or she will have more influence on your working environment than any other factor. Your manager decides what responsibilities to trust you with. Your manager will be the one demanding when deadlines need to be met. Your manager will fill out your performance evaluations, determine your salary increase, and determine your future opportunities in your immediate future.

Your project is vital because this is what you'll be spending the majority of your time working on. If you're motivated by your project, you will work well, and you'll enjoy it. When you work well, your performance gets recognized, and factor number three (salary) is positively affected. If you're not motivated by the project, it's going to be difficult to be motivated about the smaller responsibilities for the project. This leads to poor performance…you see the two different cycles. Choose a project that will place you in a cycle of success, and this will pay off best in both the momentary and long-term respects.

As for salary…The world is full of altruists proclaiming the virtues of not being concerned about money. Well, if you're content to live a life filled with poetry and poverty, then I guess it really isn't very important if you make any money. But if you enjoy things like music, traveling, cars, and have any ambitions to raise a family and clothe them in more than rags, money will be important. The most important factor in financial matters is living within your means. That being said, your means are determined by your income, which in turn determines all other financial decisions. The more you make, the more you can afford.

> Having made that comment about salary, remember that it comes in third on the priority list. *Never, ever, ever take a less interesting job for the money!!!*

Money will not erase the fact that you are bored 40 hours each week. Money should only be a determining factor between jobs that are equal in all other merits. It's important to be fairly compensated, but don't let a few bucks stop you from pursuing your dream.

Pick your boss Reiterating the point about picking your boss, we can take the advice of Stephen Oswald. Mr. Oswald spent several years as a Navy test pilot, piloted three space shuttle missions, and served as Boeing's Program Manager on the Space Shuttle Return to Flight, following the *Columbia* accident of 2003.

During a presentation a few months before the shuttle returned to space, a young engineer in the audience asked, "What advice would you give to young engineers just entering the field?" His face lit up as he started thinking quickly.

"Great Question. Pick something you like, of course, but that's what everybody says." To my surprise, from there he went on about how to play the political game, how to get paid more, and how to get onto projects you like.

"It's a funny industry. You don't get paid according to what you're worth. You get paid according to what you made last year."

His suggestions didn't have anything to do with engineering. It was all about working your way around the office and the companies. In the mix of all that, there was one suggestion he put a pointed emphasis on.

"Pick your boss. More than anything, the boss you're working for will determine how well your performance gets noticed, and your opportunities in the future."

If you've been given offers from two different companies, both of which you would enjoy doing, then you really ought to seriously evaluate who you will be working for. If the company has given you an offer, they want you on their team, and they will be willing to bend over backwards to ensure that you take their offer to come work for them. They don't want to loose you to some other company. We have actually brought candidates back through for a second visit, at their request, when they were weighing a competing offer.

If they give you an offer, they want you on the team. When you understand this, you will feel much more comfortable calling back the company, and asking to speak with the manager you will most likely be working for. You don't need to ask him what you will be doing, but ask about the team. Ask how he likes to manage the team, how the team members themselves work, and even how he would want his ideal new engineer to work for him. Don't waste his time, but get a feel for his style, so you can judge who you will be happiest working for, who will give you the type of responsibility you want, and who is structured in a way you work most effectively.

Nothing can compensate for good work, but your manager will be the one who decides where your career goes in the short term. You manager will decide to trust you with bigger projects. You can ask for all the responsibility in the company, but you won't be given any responsibility unless your manager decides to trust you with it. Your manager will determine what your raise will be, and will represent you to the managers above him. You want a manager who is on your side, and who wants you to excel in your area.

Each step follows the previous. Pick a job and a manager that will lead you in the direction you want your career to go.

Seek out opportunities for networking. These will be valuable for your own exposure, as well as developing interviewing skills and leading to real interviews. Follow up with an email the next day, and a phone call the next week. If you're still looking for a job several months later, follow up again. Remember the interviewer's perspective. He or she is a person just like you. Spend what is necessary for interviewing. The total cost of interviewing pales in comparison to your overall university education. When deciding which offer to accept, a significant consideration is to pick your boss. And never, ever take a less interesting job for the money.

Section 2—Your First Year as an Engineer

CHAPTER 7

The First Two Weeks

"You're the new guy? What's your name? Don't tell me. You'll forget my name, and then you'll feel awkward asking me my name when I remember yours. I hope they told you about the dress code. They didn't? Well, that tie you're wearing, since you didn't know we'll let you get away with it today, but if you ever wear a tie in here again, we'll cut it off and hang it above your office high enough that you can't take it down."

Your introduction to engineering may be a little more abrupt than you expect.

Most companies make a sincere effort at introducing new engineers into the company, and making that transition into the team as smooth as possible. They set up programs, have HR make every effort to communicate with managers, and do everything else that a corporation can imagine to construct the perfect introductory environment.

In spite of all these efforts to welcome new engineers and feed them into the system, there is still very little predictability as to what those first couple weeks will actually be like. Some engineers start in, and from day one, they're a contributing member of the team. Most often, there's a period of intermittent stops and starts with sporadic projects until they find their place on the team. And sometimes, engineers will find themselves starting in a company, with no leader or direction at all to help get their feet on the ground.

Whatever your experience will be, you'll make it through just fine. Millions of engineers have done it, and millions more will have to do it. They found their way around and settled into successful and enjoyable careers, and you can expect the same. One thing you can rely on is that no matter how rocky your transition may seem, the engineers you're working with have seen it before, and they're entirely confident that you'll become an outstanding engineer.

Companies generally expect new engineers to come up to speed over three months. The most important thing is to become aware of what the team is working on, and learn how to become a contributor. You do this by choosing to be interested in the work, and effectively taking advantage of the initial assignments you are given.

First, we'll go over some general Do's and Don'ts to keep in mind for any workplace. Then we'll look over some typical situations engineers found themselves in, and how they dealt with them.

Do's

- Show up on time and keep regular hours. New college graduates sometimes fail to realize the importance of this. People get frustrated if they go looking for you and can't find you.
- Be positive and seek out every opportunity to learn.
- Seek people out and talk to them. Make phone calls instead of emails, and meet in person as soon as possible.
- Ask questions. People will expect it when you first start out, and it is your best opportunity to learn how things work.
- Be personable. Get to know your coworkers and team members on a personal level instead of being strictly business.
- Learn the management structure. At a minimum it can keep you from putting your foot in your mouth.
- Use mistakes as learning opportunities. We all make them, don't repeat them. Don't try to hide your mistakes, try to learn from them.
- Be accountable. When you commit to something, do it.
- Step up and take on challenges. Initiative counts for a lot.

Don'ts

- Don't whine. Avoid cynicism, sarcasm, defensiveness, and arrogance. These will earn you a bad reputation and make working with you miserable.
- Don't take things personally. If your idea is rejected, don't brood about it.
- Don't bring your personal biases and agendas into the work place. You are there to meet the objectives of the company, not your own.
- Don't bring your personal life into your work life. If you have a legitimate problem in your personal life and you need to take time off, then let your manager know and take care of it outside work.
- Don't infringe on things like harassment policies. Remember the rule: Don't say anything at work you wouldn't say to your mom.
- Don't mix business with romance. Before you give into temptation, ask yourself just one question: "Is my job worth it?"
- Don't push the limits on the dress code. If you don't see people walking around in shorts, don't be the first to try it.

Ideal Transition

Derek Wright started at a major aerospace company in Sunnyvale, and things couldn't have gone smoother.

Day one was eight hours in orientation, and then off to the manager who had interviewed and hired him. From there he was brought to the lead engineer he would be working for, who had also interviewed him. His lead engineer introduced him to the solar array mechanisms they were working on, and that was it.

Day two was another eight hours of orientation (which continued for two weeks), and then back to his lead engineer, who brought him up to speed on the particular mechanism they wanted him to start off with. It was a hinge for a solar array to be deployed on their satellite. There's nothing glamorous about a hinge, but it's a single point of failure, so it absolutely must work. No redundancies, no alternatives, no excuses.

They had the mechanism designed, but needed to run several tests to verify the structural properties, durability, and performances of the mechanism. They asked Derek to brainstorm some ideas for how they would test for the spring constant of the hinge under transverse forces. A few days later, they met again, and went through the ideas he had generated.

This wasn't a complex demand. The testing was straightforward, and the test fixture could be made relatively easily. It wasn't rocket science, but they needed an engineer to carry out the task. Derek was responsible for designing the test equipment, working with an outside machine shop to build it, and later for actually performing the test. He worked on it for a bit each day after orientation, and then full time when orientation was done. No confusion, and no time lapses between tasks.

He had to learn to use new CAD software to model the test equipment, learn who to work with for finding the machine shop and contracting the parts, and all the other details associated with finding your way around a company. He still had many things to learn to get oriented, and to design and execute the test correctly, but his manager was fully prepared and the transition went smoothly.

This really is typical of how things go when new engineers actually get their feet on the ground and start moving. Some will have fits and starts with tasks of greater or lesser relevance, but the point comes where you get your hands on a project that matches your ability and the needs of the company sufficiently that you can take it and run. From there, your proficiency within the team increases, you become aware of what needs to be done and how you can contribute. Then you leave the phase of having nothing to do, and enter the phase of having more to do than is humanly possible.

There isn't a whole lot to learn from this, except that it proves it really is possible for your starting experience to be entirely positive. There is hope that you won't have a more typical nightmare story.

Data Crunching—A Chance for Learning

When a new engineer shows up, there are two tasks that almost everyone gets handed: data crunching and reading. As the low man on the totem pole, you're an easy target for routine data analysis. It needs to be done, nobody else wants to do it, and it's a fairly easy task for an engineer unfamiliar with the team and projects. As for reading, when you're new to the team, it takes a while to get integrated and find your place. Reading always seems like a useful fallback assignment when your manager doesn't know what to do with you at any given moment. Both of these scenarios are almost inevitable, but when handled correctly, they can be very advantageous.

The first task I had thrown at me was data crunching. My experience wasn't as smooth as Derek's ideal transition, but it was very typical of many engineers. All I knew before signing on with Northrop Grumman is that they were one of the top aerospace companies, were rated Company of the Year by Forbes, and they were hiring me to work on satellites in Southern California. As for what I would personally be doing, I didn't have much of an idea.

The Northrop Grumman facility I was at had hired several engineers and interns starting at different times that summer. When the eight hours of orientation were finished, it was time for somebody from each of our groups to pick us up, show us around, and introduce us to the teams and manager we would be working with. We gathered our papers together and were looking forward to seeing the work we'd actually be doing.

Among the people who showed up, there was a tall guy, with an attitude of casual amusement in his speech and movement. This would be my mentor.

My mentor turned out to be a pretty good guy. Having graduated from my same university, he was naturally the person to introduce me to the team (he graduated back before I even existed on this planet). Walking through the buildings with several vacuum chambers we use to simulate space and test the satellites, he casually talked about the pumps, optics, instruments, and whatever seemed relevant at the time. Then it was off to find my cubical where I would start my life as an engineer.

Mentor: "Where's your office? Building One? I've never been to Building One. This'll be interesting."

We took the elevator up to the third floor, found my cubicle, and set down my box of supplies for getting started. We hadn't talked about what I would be working on. I assumed I would be working with my mentor in some way, but none of that came up.

It was about then that my mentor told me, "Well, it looks like you're set. I don't know what we're planning on having you do. We'll probably find something for you. I'm taking off to Space Dynamics Laboratory out at Utah State in about three hours. I should probably be back in a couple weeks. We'll see. You need something to do though, don't you? Hmm. Maybe Lucy has something for you to do. You ought to talk with her."

Not recalling the name, I asked, "Did we meet Lucy?"

"No. You can give her a call though. She should be in tomorrow. Is Lucy going to be your manager?"

"I haven't seen or heard her name until just now."

"Hmm. She might be your manager. If she isn't, then I guess you can work for whoever is your manager. I think she might be your manager. If not, she can probably find something for you to do until you find out who your manager is."

"Do you have Lucy's extension?"

"Oh yeah. It's 5555. Or you can find it on the directory. Do you know how to use the directory yet? You'll figure it out. Ok? Well, good luck with everything."

And with that, my mentor walked down the hall, not to be seen for another three weeks.

I was surprised at the apparent disorganization, but at least I had a phone number. I was confident somebody would know where I belonged, so I didn't stress about it.

I called Lucy the next day. She wasn't sure if I was working for her either, but she said she'd be happy to take me until somebody else laid claim on my time. She recommended I call Brent, another engineer working for her, and see if he had anything for me to do.

None of the people I met during the rest of that week were people I had been introduced to on that initial tour. With this kind of a runaround, and an apparent ignorance that I was even showing up to work, I was wondering how this place ever got up and running. Quickly, I was beginning to think that these organizations really may not be all that organized.

I called Brent, and even though he had no idea a new engineer was coming on, he was more than happy to hand off some of his work. He worked in an entirely different building, so I had to find my way over there to meet him.

After the introduction formalities, he handed me a CD and said, "You're going to hate me, but here. We need this data processed and charted. It's temperature data from the last three thermal-vac tests."

"Ok." I paused. "What are we looking for in this data?"

"Hopefully nothing. The tests all went just fine, but we need to look back over the data just to be sure."

"Ok." Another pause. "So, give me an idea about where this thermal data came from, what these temperatures are from, what kind of patterns we are looking for...."

"Oh yeah. That's probably important, and you don't know that yet."

From there he proceeded to open up a file that had already been partially compiled, and began describing the thermal shrouds in our test chamber, their interaction with the payload under test, and what we expected to see during a routine test.

There I was, like so many engineers before me, being given my first assignment as a real engineer, analyzing data that would hopefully be totally insignificant. Fortunately, along with the assignment, I got cued into something else much more interesting and useful.

When you're given data to analyze, this is a golden opportunity to learn about the work the team does, and to begin finding your place on the team. Instead of just running the numbers, use the occasion to ask about where they come from, and what they all mean. Find out what is important in the data, and you will find what is important to the team.

In the process of describing the test chamber, Brent made the comment, "You know, it would be nice if we had a program that you could just push a button, and have all these files automatically compiled, formatted, and charted. Then we wouldn't need to bore you to death with doing all this."

The idea of writing a program to automatically process all our data was something I could take an interest in. Besides, it sounded faster than opening, formatting, and compiling files by hand for over a week (and having to do it again after the next test).

"I think I can do that."

"Yeah? If you can program something to automate this, that would be great."

And that started a whole series of analysis tools that I got to program with Visual Basic in Excel for all our routine data analysis. Before I graduated, I had never even opened the Visual Basic Editor. I had seen it once, and it looked like the C++ programming I had been exposed to in class, so I thought I could do something with it. I didn't previously have a specific interest in programming for data analysis, but seeing the opportunity in that moment, it certainly was something that I could take an interest in. I had learned the art of *Being Interested by Circumstance*, so I applied it to this situation.

Being interested by circumstance *Being Interested by Circumstance* is the art of creating interest in whatever is most beneficial for you to be interested in.

There is a difference between *something being interesting*, and *somebody being interested. Being interested* is a mental state of purpose with regard to something (a thing, person, idea, process, etc.). When you are interested, you are voluntarily conscious of something with the purpose of understanding, creating, reacting, or enjoying. You create interest in something by choosing a purpose to become conscious of it.

When you learn that interest is a function of energy input to becoming conscious of something, you learn that you can create interest in anything. That's why engineering is such a great career; you have purpose to deal with all the complex stuff in the world, so you'll never have a shortage of things to be interested in. As long as you have energy to be interested, your job can be interesting.

No matter what you're assigned to work on, you can be interested in it. Dig your mind into the interactions within the system. Figure out the science. Understand how it works, why it works, and how you can predict and control the variables within the system. Choose a purpose to make yourself conscious of as much as you can regarding the *something* you are *being interested* in.

You will have to put energy into understanding. You will have to put energy into creating ideas. You will have to put energy into checking if your ideas are correct. Bottom line, you will have to put in energy. But if you do this, the project you are ripping into will become interesting. It may have not been interesting before, but it will become interesting.

When you drop in with your team, and they're not sure what they need you to do, keep your eyes and ears open to the needs they have. Any project always has more needs than it has people to address those needs. The most critical needs are the ones that people take an interest in first.

When you're at the bottom of the seniority stack, you get to pick from the lesser priorities to decide what you will be interested in. Your circumstances determine what interests are available to you within your job. Acknowledge your circumstances, and choose the most important thing for you to be interested in. Once you pick a problem to be interested in, run with it. Wrap your mind around it, create understanding and solutions, and you will find yourself becoming a valuable contributor.

Right now, you may not be very interested in thermal analysis, stress and strain in plastics, high cycle fatigue in switches, or any of the millions of things you could end up working on. But when you find yourself on a team working on these things, you can acknowledge your circumstances, and choose to be interested in them. When you make that choice, it might surprise you, but you actually will become interested. When you're interested in what you're working on, you're bound to enjoy it. It may not be glamorous, but it will be interesting.

"Read This"—Learning Documentation

All managers know that reading can be dull, boring, and entirely uninteresting to a new engineer. Seldom will reading be the first assignment they give you. It's a last resort for when they have no idea what else to do with you. It probably won't come up until you have completed a couple simple tasks.

"Here are those charts you wanted."

"Great. I'm really glad. Those will be useful."

"Now what do you want me to do?"

"...uh...hmm...."

Now your manager begins sifting through notebooks in her office.

"Here, read this," as you feel the weight of an inch-thick document drop into your hands. "...And this is useful," while handing you two more notebooks. "...Hmm...this is probably too detailed to be of any use...but this would be good to know," and another couple pounds of ink and paper are abruptly added to your burden.

"Good. That should keep you busy for a while. I know you don't want to just read, but this will be very useful to learn. I'll let you know when we have something else."

You put on a smile, glance at the books, struggle to refrain from flinching, look back at your manager, and say, "Ok. I'll be at my cube."

That may sound a little dramatized, but thousands of engineers have had that exact same conversation. It's more common than anybody would like it to be.

This is where you need to begin understanding what your manager really wants. A conversation like this doesn't really mean your manager wants you to read all those documents. What was said is actually manager code language for two things.

First, "I don't know what I can have you do, that you know enough to do."

That's what "...uh..." means.

Second, "Learn what we're doing around here, so you understand enough to be useful."

That's what "Read this" means.

It's nothing personal. It's just a factual recognition that there is a learning curve involved with becoming part of an engineering team.

In this situation, where your assignment is blatantly incredulous, it is important that you understand the real need of your manager. When you're handed a five-inch stack of documentation that covers every generality and detail of the project your team is on, your manager doesn't have any expectation of your assimilating all that information. The real goal your manager wants is for you to become functionally familiar with the project. This is the goal you will keep in mind when you take these documents back to your cube.

When you sit down, instead of being intimidated or overwhelmed, you can be interested by circumstance. Circumstance has dropped a stack of technical documents in your hands, so choose a purpose to become conscious of what is in them. You already know enough material exists in there for you to be interested for months on end. You won't read from front to back. You will hop around from interest to interest, building an understanding of the project. Do this, and the few hours you have with these documents will be time well spent.

Take the documents and begin scouring them for understanding, but keep in mind what your job is. Study those documents with your mind open to what your team and manager are working on. Learn the acronyms. Focus on the components and functions you have heard mentioned in meetings, or seen in diagrams on your manager's wall. Learn the overall picture, and then focus on the area most familiar to your team. This will help you to learn the language your team is speaking, and enable you to understand and contribute constructively to discussions.

While you are up to your elbows in these documents, and it crosses your mind that you want to design, not study, realize that the faster you learn these documents, the sooner you will be positioned to contribute as a team player. The more you learn from these documents, the more areas you will be able to get your hands involved with relating to the project. It won't be long until your manager has something else useful for you, so take this time to develop as much proficiency as you can, and it will open doors for you in the future.

Sharon Redd started off with a team working on integrating the JDAM (Joint Direct Attack Munitions) bombing system, commonly referred to as "smart bombs," into the Air Force's supersonic bomber, the B-1. These smart bombs are the same used in the F-117 Stealth Fighter, delivering precision guided destruction right down through the air vents of their targets. The B-1 is supersonic, whereas the design of the Stealth limits it to subsonic flight. The smart bombs were developed later than the B-1, so a team of engineers was needed to modify the aircraft for using this new weapon.

Weapons integration is a challenge that has grown exponentially with advancements in technology. We used to only need smart engineers to develop the weapons. Now we need smart engineers to make all these weapons work together. This is typical of the affects of technology in any industry. A car used to be an engine with wheels and seats. Now we are integrating GPS navigation, proximity sensors, traction control systems, and so on.

Integrating new technology is not a simple thing, but it must be done to remain competitive. This in particular is a superb place for new engineers to work with experienced engineers. The new technology must be learned, which new engineers are readily adept at doing. The old technology and systems must be understood, which experienced engineers have already been doing. Both of these must come together to integrate new technology with existing systems.

When Sharon started with the team, she was handed the customary stack of documents and instructed, "Read this." She walked back to her office with several pounds of documentation on the old B-1 weapons system, and the new JDAM system. Sharon was smart about the situation, so she didn't *read* the documents, she *learned* the documents, and *learned* the different weapons systems. She took an overview of what the documents covered, and began building an understanding from the ground up. She went from one interesting section to another interesting section, finding answers to the questions that naturally arose about how everything works.

How do the pilots guide the bombs? How does the B-1 navigation system communicate with the weapons system? How do the pilots drop the bombs? Unexpected fact: There are different flying maneuvers for dropping the bombs. How do these different maneuvers affect releasing the bombs?

Sharon's questions didn't come to mind in the same order that the documents were organized. That's why she didn't *read* them from front to back. She *learned* them from point of interest to point of interest. If we try to read from front to back, 99% of us will fall asleep every time. But if we learn by moving between points of interest, it becomes exploration, and is genuinely interesting (and significantly more effective).

Sharon's team was working on the communication between the two systems, which operated on different protocols. Unfortunately, making two systems communicate is more complex than simply translating from one language to another. Signals must be reconfigured for different interfaces, the fundamental command structures don't coincide, timing issues arise, and so on.

When they mapped out the overall systems they were going to integrate, Sharon had already become familiar with these major components. When they dug down into the details of how they were going to manage the data flow back and forth, Sharon was familiar with where to find the needed information in the documents. She didn't remember the details, and hadn't read many of the details, but she knew where to find them. By understanding the Big Picture of what they were working on, understanding the medium picture of their particular system, and knowing where to find the details of their system, Sharon quickly became a significant player on the team and thoroughly enjoyed working on the project.

Surfing Meetings

When you join your new team, you will have to be an inventor in order to create your role. The principle of invention, the very basic definition and thesis of all invention, is making something useful that didn't exist previously. Invention is the label we attach to the process of useful creation.

People typically think of invention as devising some new useful contraption, a new chemical compound, or a genius algorithm to compute things faster. In this context, we're speaking of invention as creating something useful to your team. If you want to enjoy your first six months, you will have to be an inventor. You must invent something useful for the time you have available to work.

No matter what your starting circumstances, if you put your mind to it, you will be able to find a role that you will enjoy, and that will be valuable to your company. Most likely, you will have a manager and team to work within, and will invent solutions to their needs. But even if you don't have this foundation, if you put in the effort, you can find useful work.

Curtis Broadbent started at a large company with a degree in physics, a bright mind, and an interest in their industry. What he didn't have was a manager claiming responsibility for him. He had to be a little more creative about finding his place in the company, but he stepped up to the challenge, and found a position that he really enjoyed and that was valuable to the company.

After the initial day of orientation, it took another couple days for him and the people he called to realize that he hadn't been placed anywhere. With a physics degree, and no manager, there really wasn't a predefined function for him to fill. Being assigned nowhere opened the opportunity for him to go anywhere.

He contacted the managers over different teams, and asked if he could attend their meetings. The managers were happy to have him come, and it served for him to develop a broad view of what was going on with many different aspects of the company. Furthermore, he was able to meet many people within the company doing all sorts of work.

In addition to attending meetings, Curtis actually *asked* for project documents and reports to read. He didn't even have the luxury of being given the reports, but he knew what to do with them.

As the meetings would go forward, there were always tasks that needed to be done. Curtis volunteered to pick up the tasks that were interesting and suited to his ability. He spent a couple weeks with this routine, networking to other teams, and then narrowing down to teams where he naturally was more interested and valuable.

Eventually, he gathered enough understanding of the teams and what they were working on that he saw some cross-team needs for coordinating analysis and other aspects of their responsibilities. Because he had investigated what so many teams were working on, and had studied so many aspects of the projects through the documentation, he was functionally aware of several teams' needs. He became valuable in a cross-team role that the teams had previously been unable to perform.

Curtis wasn't given his position; he created his position. He could have sat at his cubicle, complaining that the company didn't have anything useful for him to do. Instead, he took an interest in the work going on, put in the work of studying the project documentation, found needs, and invented solutions. He started off in the most inopportune position possible, and created a job that he enjoyed, and that several managers appreciated.

Creating substance instead of fluff Curtis is a prime example of *Creating Substance Instead of Fluff*. He had zero hours of work given to him each day. He created 40 hours of productivity. When you start on your team, for the first few months, it is likely that you'll only be given 30 hours of real work to perform. That leaves 10 hours empty. You decide what to fill those hours with.

Whether you like it or not, you will have to be an inventor in one of two ways: you invent substance, or you invent fluff. If you invent ways to stretch your 30 hours of work to fill 40, filling the time with nothing, you are creating fluff. If you do this, your invention is only useful to drag out your misery at work.

Your other option is to become an inventor in the true meaning of the word. You do your 30 hours of work as efficiently and thoroughly as possible, and use the rest of the time you have to create something useful.

Invention is the key to making substance instead of fluff. Your manager has already indicated she doesn't have anything else to give you, otherwise she wouldn't have given you documents to read. You can't go to your manager for more work, so you'll have to find it on your own. Studying the documentation is a good start, but you want to get your hands involved in some work.

You find more work to do by taking an involved interest in the work going on. If you were given data to analyze, then you have an excellent excuse to learn more about where the data came from, and what they mean. Other engineers on your team will be working on their projects. Talk with them about what they're working on, what influences their design decisions, why particular analysis is useful, etc. As you begin expanding your understanding of the work your team is involved in, it will become apparent where there are needs you can work on.

You've talked with other engineers on the team, and learned that one of the engineers could use some help with the analysis on an aspect of his project. He's already overworked, but while you were asking really good questions about the work he's doing, he saw that you're smart enough to help out with a part of it he hasn't been able to get to yet.

Great! You just found your next assignment. You're happy about it, the engineer you're helping is happy about it, and your manager will be happy about it when you inform her.

You don't have to go to your manager to find everything you're going to work on, but your manager does need to know what you are working on. She hasn't forgotten about you. Integrating you into the team is one of her priorities.

Your manager is interested in moving you beyond the introductory transition, and getting you involved with your own projects. In order to do this, she needs to see what you're interested in. From your side, you need to become sufficiently aware of the project that you can be interested in it. You can't be interested in something if you're not conscious of it.

On these initial assignments, it is very important to follow through and meet your commitments. You want to establish a pattern of reliability. By working efficiently and effectively on the work you have, and then searching to invent more useful work, you speed the process of becoming an integral part of the team.

Remembering Names

From Dale Carnegie, we learn that the sweetest sound in the human language to any individual is his or her own name. People love to hear their name. It offers recognition, commands attention, and distinguishes one individual from another. Calling somebody by his or her name is the most effective compliment you can offer, and is an essential building block to developing trust with people.

Recognizing this fact, it is important that you remember the names of the people you meet during your initial weeks on the job. They won't be offended if you forget a couple times, but they'll be sincerely impressed and appreciative if you remember the first time.

Remembering a person's name also makes it 10 times easier to ask a quick question if you don't have to precede your question with "What's your name again?" You'll be asking a lot of questions your first few weeks. Hopefully names won't be among the set of questions.

Remembering names is a skill. Many people feel that they're terrible at remembering names. This isn't because they're incompetent; it's because they don't practice the skill frequently. Ask yourself, when is the last time you made a concentrated effort at developing the ability to remember names? Even if you have a terrible track record with names, you will be amazed at how well you can do with names when you actually concentrate and focus on it.

Here is a simple tool to help you along:

1) Repeat
2) Relate
3) Recall
4) Write
5) Review

1) Repeat When you are introduced to somebody, repeat his or her name while shaking hands. Don't just parrot the sound you heard, but consciously decide that you are registering this person's name into your mind. It's sort of like mentally pressing the *save button* with this newly entered variable you're associating with this person's face and voice. Be aware that you intend to remember this name you are saying, and that this name belongs to this person you are meeting.

2) Relate This is like finding the file you are going to save the name of this person in. What you are going to do is save the person under a file you already have built for that name. Relate the new person to a person you already know by the same name. We've all met at least five Dan's in our life. Find some way to relate this person to another you already know by the same name, by a feature of his or her looks, trait of his or her behavior, or any characteristic. Later, when you're trying to recall the name, whatever feature or trait you related will easily come to mind, reminding you of the person whose name you can already quickly recall.

3) Recall Before you conclude your conversation with the person, consciously recall the person's name. Using the same trait relation you already programmed in your mind, bring the name back to your conscience, and conclude the conversation using the name. "It's good to meet you Natalie."

4) Write It isn't difficult to scribble a name along the edge of a paper you have on hand. When you have a moment, write the person's name down. This is another conscious act of confirming the name of the person you just met. It also gives you a small reminder when the day ends and you have met many people.

5) Review At the end of the day, take five minutes to consciously go through the names and faces of the people you met. Gather the notes you scribbled names on, and make a cheat sheet with the names of everybody you met that day. Consciously recall their faces, and the relations you made for them. Run through these name-to-face connections until you can name everybody without using your cheat sheet. Keep that cheat sheet around because you'll want to run through it first thing when you arrive in the morning. It is a small time investment, it requires a small amount of conscious thought, and the payoff will certainly be worthwhile.

Know the people you work with. Know their names. Doing this will increase your confidence in them, and increase their confidence in you. If you pay attention to nothing else on that first day, pay attention to the names of the people you meet. If you remember nothing else from your first day, remember the names of the people you meet. Again, they won't be offended if you forget a few times, but to remember is certainly possible, and very worthwhile.

There's no predicting what will happen during your first two weeks. If you follow a few do's and don'ts, it will help you avoid the catastrophic mistakes. You will probably be given data analysis. Use this as an opportunity to learn the projects the team is working on. You will probably be given stacks of documentation. Instead of *reading* these documents, spend your time *learning the documents*. You will periodically have extra time on your hands. Be an inventor, and create something useful with this excess time. Remember the names of people you meet, and make every effort to meet people in person the first time, or at least make a phone call. Avoid email until you're familiar with your coworkers.

CHAPTER 8

Finding Your Place—*Skills for Typical First Projects*

Sometime along the first few months, you wind up on your own project, and begin finding your place on the team. It could happen early, or it could happen later, there really is no predictability. Let's look at some principles and tactics useful for getting yourself into these projects, and working through them.

Be an Engineer

Derek Wright, whom we met earlier when we talked about his start at a company building satellites, finished the initial project they gave him when he walked through the door. Then, like all new engineers, he had to find his place within the rest of the team. He started working with another engineer who was having difficulty with a test bed he had been assigned to develop for an antenna pointing mechanism on a satellite.

This particular satellite was being designed for an 18-year life cycle. As with everything on a satellite, you can't send somebody into space with a wrench to fix things that break. Everything must be designed and tested to ensure that it will work for the lifetime of the system.

They created a test bed with two stepper motors that needed to be operated independently and simultaneously. In this setup, a glitch was occurring every time they flipped the switch to make both motors operate simultaneously. One of the motors had a delay, and the operation counter would jog several counts. The original engineer had found a series of seven commands that he included in the operating procedure to get around this problem, but that didn't fix it. They still didn't understand what was causing the anomaly, and they were afraid the glitch would show up in other places during the test.

As it turns out, this other engineer just wasn't at all interested in this test bed. He was a mechanism design guy, with no interest in making computers talk to hardware, or anything electrical. When Derek came onto the scene and started digging into the electronics, the original engineer was more than happy to hand off the project.

Derek picked up the system, figured out what was causing the problem, and fixed it. Then they could use the test equipment the way they intended to use it, and didn't have any more unexpected results from commands they sent the system.

Once you get involved with a project, and do a good job with it, you usually end up following it all the way through. They wanted to test the full 18-year life cycle, and they wanted to run that test as quickly as possible. Derek completed the test bed, and then went on to design an optimal test path for the antenna pointing mechanism.

When you find yourself on a project like Derek, and the system is almost working, but not working right, you just have to *Be an Engineer.* You know you're smart. You know the system operates on logic and the unchanging laws of physics. You know the system can be understood. If it can be understood, then it can be designed right. You know this. All that's left is for you to break out your technical mind, dig in, understand the system, and fix it. You just have to dig in and do it. Your mind can handle it. *Be an Engineer.*

This is the trait that makes Derek Wright such an effective engineer. It never crosses his mind that he might not be able to understand something. Whether it's making computers talk to servo motors, learning what somebody else is working on, or writing equations in the sand to calculate the height of a waterfall he just threw a rock from, he's confident he'll be able to figure it out. He doesn't stop and wonder if he's smart enough. He doesn't stop and wonder if the challenge is too complex. He just starts understanding it.

Too often, people quit trying to understand as soon as they see something about which they have no current knowledge. It's like the first time you looked at a calculus book. It can be intimidating in its unknown complexity. The nature of an engineer is to face up to the complexity, and to make sense out of it.

This is what you spent the last four years doing. You have four years of experience in the art of technical learning. That is why your company hired you. This is why you are called an Engineer. When you see something new and complex, you pick your head up, rev up your mind, and go at it. It may not be easy, but you can understand it.

As engineers, we don't need to worry if we're smart enough. We just need to do it. If you're considering skiing a steep rocky mountain, you really need to think if you're good enough to ski it. If you're not good enough, it could kill you. But as an engineer, there is no danger whatsoever in getting halfway into understanding something and finding out it's incredibly complex. Who's going to get hurt? Nobody. Rocks kill people; understanding doesn't. When is the last time you read about an engineer who was found dead at his desk because he thought too hard about a problem? Are you afraid an integral is going to come out of nowhere and give you a concussion? Engineering is totally safe. Don't hesitate about it, just understand it.

Be an Engineer!!!

(Note that this is in reference to *Understanding*, not actual *Implementation.* In the section "Moving Too Fast" in the chapter "Guessing What your Manager Wants," the prudence necessary for engineering within complex systems is addressed. *Understanding* won't hurt anybody, but *Implementation* can become dangerous and very expensive in a larger system.)

Choosing your career interest Another principle to look at in this example is the way you direct your career by choosing what to be interested in. After you get integrated with your team, and you have more work than you have time to do, you will find that you have begun developing an expertise within your team. You'll become the go-to person for your own carved out niche.

This is the time for you to start being more selective about where you want to take your experience. The engineer who was originally assigned to the project Derek took over was not interested in that nature of work. He was a mechanism design guy, and that's what he wanted to do. He was developing this test bed because it needed to be done. Part of being a team player is sometimes doing things just because it needs to be done.

When the opportunity came for him to transfer the project to another engineer who was interested in it, he made the hand-off. The project was getting done, and he was free to spend his time working on projects more in line with his long-term career interest.

Contributing Solutions

One of the most effective ways to get involved in a project is by discussing and suggesting solutions to the technical challenges through which the team is working.

Mike Mamanakis dropped into our team and started off like a rocket. A quick mind, he was looking to learn new things, and was very diligent in carrying out the initial project we assigned him. In addition to what he was assigned, he kept his head up about everything else going on with the team. He didn't wait around for us to tell him about everything in our facility, he asked questions, and rapidly developed a detailed understanding. In short, he was the ideal platinum-grade engineer.

The analysis we gave Mike to start with was directed toward understanding a particular anomaly we had experienced during our most recent test. Instead of just crunching the numbers, making charts, and asking what format we wanted them in, he asked about the nature of the problem itself. By understanding the problem, it became apparent to him what analysis would be best, and how it would be most effective to present the data. While understanding why and how the data he was analyzing were relevant, he became constructively involved in our conversations about the cryo-cooling plates we thought were the root cause of this problem.

In our test chamber, we have a couple liquid helium cooled thermal radiating plates that we use for simulating the radiation heat loss our satellites experience on orbit. We had recently dismantled these plates to remove the thermal grease that had been used, and to replace it with indium foil. Indium is a very soft metal that offers excellent thermal conduction between two surfaces. These surfaces each cover eight square feet. Thermal conduction is highly influenced by surface contact, so surface variations over this large of an area can greatly impede the overall heat transfer.

We needed to check if the plates were flat enough that a 0.005″ layer of indium would give us the contact we needed, without large gaps. The plates themselves were warped, but they had a closely spaced bolt pattern. The constraining factor was that we only had two weeks to have the whole thing dismantled, fixed, and reassembled for the next test.

It was by taking an interest in this project that Mike really got his feet on the ground and started moving. He didn't wait for us to tell him to search out solutions to these challenges. He jumped right in as part of the team, discussed the problems, and found solutions.

Sometimes new engineers feel intimidated among more experienced engineers. They feel their experience is limited, and their ideas aren't as valid. Sometimes, they feel they don't understand everything in a system, so they're not in a position to suggest solutions. This mentality is understandable, but it couldn't be further from the truth. You have an engineering mind, and the sooner you start applying it, the more benefit you will bring to the team.

Mike never hesitated to be an engineer. I don't know if it ever occurred to him that he was a "new engineer." Right from the start, he dug into understanding the systems, our objectives, and discussing solutions. There was a lot he didn't understand initially, but through asking questions and offering solutions, he quickly learned how the cryo-plates work, and all their considerations regarding the payload.

While discussing how we could measure the gap between the surfaces, Mike remembered a gap measurement tool he had used working on his car, plastic strips that deform a gauged distance when squeezed between two surfaces. None of us had heard of these plastic strips before, but they were just what we we needed. Mike did a Google search on them that afternoon, and we ordered them right away.

Mike had brought to us the tools we needed. Fast, accurate, inexpensive, and simple. They verified one plate was adequately flat within 0.001″, and exposed a square-foot gap of a few thousandths on the other plate for which we were able to compensate. Less than one week on the job, and Mike had already become a significant contributor.

Mike followed this project the rest of the way through. The initial project we had given him only had an indirect connection into this more involved project. It is worthwhile to point out that his involvement rapidly came as a direct result of Mike taking an interest in really understanding our projects. He wasn't passively waiting to be given an important assignment when this came up. It was the natural result of an inquisitive mind and a team seeking a solution.

Countless principles can be learned from this example, but we'll just pick a few.

First, learn everything you can, everywhere you go. It just may come in handy when you least expect it. Mike never imagined that working on his car in high school would lead to a solution for equipment in a thermal-vac chamber for testing satellites.

It is also worth pointing out again how Mike went about learning from the data analysis we gave him. He didn't focus on the charts, he focused on the problem, the reason for the analysis, instead of the analysis itself. He didn't ask questions about frivolous formatting and presentation. He inquired about where the data came from and what they meant. The details about charting the data became apparent once he understood the reasons for the analysis.

Finally, he exemplifies what it means to *Be an Engineer*. Having only three days on the job didn't deter Mike from suggesting solutions to the problems with the equipment. He never was concerned that he might not understand the equipment or help with a solution. He was continually engaged in understanding.

Finding People with Information

One of the most valuable things you can learn for working within an organization is how to track down the right people with the information you need. With the complexity of projects, all their phases of development, and all the support equipment necessary to their production and testing, there is no such thing as one person who has all the answers. The answers you need will be found from a network of individuals.

The common thread between all the projects will be finding who has what information, who knows the functional specifications, who has existing designs completed, and who knows what can and cannot be redesigned. Note that all those questions begin with "Who."

A new intern came in for the summer and started asking questions about who designs our system control racks. They're fairly impressive looking pieces of equipment. Eight feet tall, and stacked with blinking lights, digital readouts, control buttons, switches, dials, and everything else you would imagine in a high tech test facility. Just one standing alone looks pretty complex, but a whole wall made up of these racks becomes intimidating.

Somebody has to design every panel, every controller, light, button, connector, and component, right down to every wire inside of those racks. Somebody has to determine which gauge of wire to use where, which connectors to wire in parallel, in series, with which voltage, and so forth. It's a tedious job, and even though it's not rocket science, it still has to be done right.

In addition to their internal operation, the racks need to talk to each other, to test equipment, to individual computers, and to computer networks. Somebody needs to understand the functionality of the connecting racks and components, which signals are sent and received, and which wires send and receive those signals. Somebody needs to learn all this to design the actual interface lines. This would be a fitting job for...a first year engineer.

A few months on the job, you've begun to find your place in the team, when in a meeting they start discussing how they're going to integrate the TCR (Thermal Control Rack) with the SCADA Rack (Supervisory Control and Data Acquisition). The TCR is being designed and built by another team, who is more involved with the performance specifications of the payload components the rack will control. The SCADA Rack is (supposedly) being designed and built in cooperation with the team designing the TCR. You're the new engineer, so they assign you to design the interface between these two racks.

This would be a typical assignment in nature and complexity given to any new engineer. First of all, it requires quickly learning a good deal of technical material that you'll immediately apply, and then never use again. You've been doing that for the past four years (see your homework).

Second, it starts building a foundation of technical understanding that will be beneficial to you throughout the rest of your career. You get your hands dirty in working with ground level components and you develop a functional understanding of the technicalities involved with building custom equipment for any given operation.

Most important, you learn to work your way around the company, and track down people with knowledge about the system.

Many projects are similar in nature to this interface between these two racks. You may start with coordinating designs of components in consumer products to be sold by your company, components for manufacturing automation, field applications involving your company's products, and so forth. All of this requires engineering. Right now, we will roll through some of what you may experience from the standpoint of the actual design process, and the people involved in that process.

I was the new guy when this interface came up in a meeting, so it was suggested that I pick up the interface between the TCR and SCADA Rack.

The guidance they gave me to get started: "Contact Ray Luntz. He's been working on them, and should know everything you'll need."

At this time, it still wasn't clear in my mind what SCADA and TCR stood for. Those were just the acronyms I needed to use to talk about these things. Far from knowing the names, I hardly knew what they would be used for. What I did know was that they needed to interface with each other, they would be used at the chamber where we were making several upgrades, and this guy whose name I just wrote down should know everything else.

I contacted Ray the next day; he was an electrical engineer who ended up being one of the easiest guys to work with. He knows his technical work very well, but is so casual about it, you feel like you're talking about something as simple as Legos. I learned a lot of science and equations in my electrical engineering class, but I learned more about practical electrical engineering from Ray than I ever anticipated needing.

At the time, Ray was up to his elbows in the TCR. This rack was going to control the heaters and nitrogen flow to the payload during testing. He had been working with the payload thermal engineers, and the requirements for the TCR equipment had been defined.

Ray knew everything we needed to know about the TCR. Then I proceeded to ask him about the SCADA Rack.

"I think you guys are designing that. There's the rack over there in the corner, and there are several parts for it in those boxes, but nobody has said anything about what it does."

I contacted Brent Nielsen on our team. He's the engineer who does the job of three engineers. He works quickly, and knows the hardware so well that it's faster for him to fix things than to describe the problem to somebody else. While others hold meetings to decide if they have a problem, Brent is out there fixing it. Brent had half the SCADA Rack designed, but had been told by the payload thermal engineers that they wanted to design the half that interfaces with the TCR.

I contacted the payload thermal engineers. They referred me back to Ray.

In the meantime, Ray had asked another thermal engineer about the interface to the SCADA Rack, and he was instructed to work with Brent.

This whole loop of leaving messages and getting messages returned took about three days. Each person I talked with had some kind of a vague idea of how the racks were to interface, but nothing concrete. Brent had already ordered several controllers and other components he assumed would be needed, anticipating there would be delays and misunderstandings of this nature.

Finally, I got both Ray and Brent into my office, cleared off my white board, and drew a simple block diagram with two boxes and three arrows.

When I finished, Brent identified the status of this project by saying, "That's the most comprehensive drawing of this system that has ever existed in this company."

I was shocked. By the time I talked to the third person, and certainly by the fourth, maybe I should have begun to realize that the design probably didn't exist.

Since the design didn't exist, it became my job to create it. More often than new engineers expect, they find themselves in this position. Responsible for a design that doesn't exist, and working on a system with which they have no familiarity. Now, instead of contacting people to find a design, I was contacting people to find requirements, and expertise for designing an electrical system.

I worked with Brent for understanding our facility requirements, how the designed half of the SCADA rack works, and how we wanted to operate the rack. I worked with the thermal engineers to determine how many circuits they needed for their heaters, and the performance and safety requirements of those circuits. I worked with Ray to understand how all the electronics work, how the TCR works, and how to go about designing for those requirements.

When you are a relatively new engineer, you have a distinct advantage when contacting people and tracking down information. Everybody in an office is happily willing to help out a new engineer. They all assume you're smart and talented, but are sympathetic to the fact that you don't know anybody, and are just getting your feet wet. They want to make your job easier, help you find the information you need, and are willing to teach you anything relevant to your responsibilities.

Also, don't be nervous to ask people to slow down or clarify their descriptions of how things work. It is more important for you to understand than to come across as smart.

This principle holds true all through your career, not just your first year. Understanding is why we communicate, so be sure you understand.

Don't be afraid to contact people. When it's suggested that you contact somebody, write down his number and give him a call. If he's not at his office, leave a message. You can email people, but for a first contact, always take the time to make a phone call. If you get an opportunity to meet the person at his office, take it. Making personal contacts within your company makes it a more cohesive place to work.

Throughout your whole career, you will always be crossing paths, sometimes every day, with people you don't know. When it's suggested you contact somebody, it's very likely this person you're contacting is one of the anonymous people you see every day. Having a reason to contact him is a great way to put a face and voice to his name. He'll appreciate meeting you, the information you need to discuss will be communicated more effectively, and you'll have one more friend at work.

The Efficiency of Personal Interest

We've already looked at the necessity of tracking down people for projects, but now we're going to take a look at an area where an added efficiency can come through personal contact.

One of the most rewarding projects you are likely to be assigned within a few months of starting will be carrying a project from design all the way through fabrication and integration. You get the functional requirements, you'll discuss the design intent with several other engineers, and you get to take it from there.

If you work in a large corporation, you will find there is a "wall" between engineering and manufacturing. You send the designs for your parts to the machine shop, and they make them. You don't get to work on your parts, and they don't get to design them. Nobody likes this wall, everybody knows it's inefficient for effective engineering, but it's there.

Even though a wall exists between engineering and manufacturing, with a little effort, you will find this wall has doors that are open to anybody who wants to walk through them. It doesn't take much time or effort, but showing a little bit of personal interest in your project goes a long way for getting it manufactured right (not to mention it's fun to swing by a machine shop).

Everybody is more interested in a project if there is a person associated with that project. If there is no person, it's only a paperwork project. If people know you are interested in your project, and they know you, they will be more likely to do good work on it, and get it done on time.

You have two paths for getting your project fabricated. Your company may have an internal machine shop for these things. If your internal shop is busy, or if you don't have one, you will go to an outside machine shop.

I was given the responsibility for designing an infrared blackbody to be used with our optical sources. Our radiometric Ph.D. who guided me on the design principles suggested that I call the outside shop we were gong to have make the part, and ask if I could stop by to discuss the part the morning after we emailed them the drawings. The shop was happy to have me come by for a visit.

When I arrived that morning, they had gone over the drawings and had written several notes on them. I ended up talking with the project manager, the guy programming the CNC lathe, and a couple shop machinists. This part had a unique face, and their machinists' ability was a driving factor in how this would be created.

They had questions about all the things I thought they'd have questions about, and we couldn't have cleared those up so clearly through email or over the phone.

If the visit hadn't been made, they may not have talked about it for a few days, and it probably would have been a few days more before they got back with our procurement guy, who would then get to my manager, then to me, adding several days to the communication process. As it was, we all got to understand everybody's design concerns, and the right decisions were apparent, so they started rolling forward on it right away. I didn't have to ask them to get a move on it, because they were already moving on it. I would estimate that this single half-hour visit on the way to work cut about two weeks off the total turn-around time.

Learning to Do Great Engineering

A tendency for many new engineers is to feel like the best way to contribute to the company and the projects is to get their work done quickly and efficiently. While this is true, there is another purpose the company wants to accomplish during the first couple years.

The primary purpose during the first two years for an engineer is to learn how to do really great engineering. Learning to do work correctly is more important than learning to do work quickly. Understanding a design is more important than modeling or analyzing a design.

As a new engineer, you will do much of the grunt work for seasoned engineers who have developed several areas of expertise. It is important to get the work done, but it is equally important that you understand the work you are doing. When you have the opportunity, take a few minutes to ask questions and learn from the expert engineers you are working with.

Developing Expertise

Many engineers are (justifiably) afraid of developing an expertise and then being pigeon-holed into that narrow area of expertise. Their interests are more diverse than doing a single type of analysis, or programming with a single language.

Some engineers get pigeon-holed. However, it's not *having an expertise* that leaves them stuck in the same place. It's having *no additional expertise* that leaves them stuck in the same place. As soon as you stop learning, you become stuck doing the same things you have done previously. You can't help but stay in the same place if you don't learn how to do anything else.

Taking this to the positive end of the spectrum, it is having multiple areas of expertise that allows you to get involved in more aspects of a project, and take on increasing responsibility. All chief engineers or executive managers need to be an expert in multiple areas of the projects they are responsible for. They don't have the time to perform the work on every aspect of the projects, but they must understand every aspect enough that they can make good decisions. If they don't have an adequate understanding, they'll swiftly lead the project or the company straight into the ground.

To manage several experts, it is essential that you personally understand what it is to be an expert. You don't have to be an expert in each area, but you must know how to become an expert, and how to communicate with your experts. Furthermore, all engineering is interrelated. The more you learn about one area of expertise, the more understanding you will have with which to relate to other areas. If you are going to develop multiple areas of expertise, you must start by developing one.

At some point, you may find yourself in front of a computer doing boring and monotonous work. Instead of contemplating the fact that you don't want to do this the rest of your life, focus on learning to do it *precisely and proficiently*. Recognize that *precisely and proficiently* are different from *hastily and quickly*. If you learn to do the tasks well, then you won't be stuck doing them the rest of your life.

We always laugh when we hear somebody say, "He got too good at his job, so they promoted him out of it." There is truth to that. If you don't want to stay in your present job, then do it exceptionally well, and you'll be promoted out of it. The sooner you develop the expertise, the sooner you'll be positioned to move on to other interests. Seek opportunities to follow the work you do into other areas of application. In engineering, everything relates to everything else. Volunteer to help with the broader applications of your work, and you will find opportunities to develop greater understanding and work on more interesting projects.

The speeches your professors give about *perpetual learning* are the key to not getting pigeon-holed. Remember their encouragement to keep learning, offer to do work that will increase your exposure, and you'll have an interesting and successful career.

While finding your place on the team, *Be an Engineer!!!* Don't hesitate to understand something. Integrate yourself into the team by contributing solutions to the projects they have under way. People are your most valuable resource. Learn how to track down the people with the information you need. Show a personal interest in your projects. Don't just assume other people will understand what, why, and how you want your project done. The first year is the time to learn to do great engineering, so take the time to understand why design decisions are made. Develop an expertise, and then develop another expertise, and keep going for as far as the projects reach. This will keep you from getting pigeon-holed, and keep your job interesting.

CHAPTER 9

Guessing What Your Manager Wants

When we first step through the door, it is easier if we are micro-managed. We don't know what's going on, so it's easiest if we're given lots of detailed instruction. But as we gain an understanding of the team and the projects, we feel crowded if we're micro-managed.

Making this transition from being micro-managed (and wanting to be micro-managed), to working as a team player, can be a point of frustration. A little understanding can help smooth the transition over that first year, so we'll address the dynamics and some skills to make it through.

When you drop into your team, you really don't have much of a clue about what is going on and how you can help. Ideally, your manager will have an initial task for you to do, that task will be clearly defined, and it will be easy for you to understand what the end product will be. Seldom does reality match up with the ideal. That first project may or may not be well defined (it may not even exist), but the chances of your second and subsequent projects being well defined exponentially decrease.

When you first show up, your manager has known that you will be coming, and he will have dedicated some time and thought to making your initial experience as positive as possible. This means providing something for you to do that seems important, that is not mindlessly easy, but that somebody who knows nothing of the team or of the projects can accomplish. Finding such a project is more difficult than new hires and interns would guess. But usually, with some creativity, a project is found, and it is held at bay for a week or so until the new hire shows up and it can be given to him.

Asking the Right Questions

Now you're the new engineer, a not-so-demanding project is in your hands, and you're supposed to know what to do with it. You want to start asking questions. You have questions for your manager, but you don't want to bother him with meticulous details. There are effective questions and less effective questions. You want to get to the effective ones in order to get your work done.

Every question is focused around two things. First, and most obvious, is the content of the question: what, why, how, when, etc. Second is *who* the question is directed toward. Your initial tendency is to have questions directed toward your manager. When it's time for you to work on your own, it can be very helpful if you start directing questions toward yourself.

This is where we begin the scientific method called *Guessing What Your Manager Wants.*

Joe Denny joined us for a summer internship, and we had some data from a previous test for him to analyze. I had shown him how to use a macro for the data, and was ready to send him off to run through the files.

On his way out of my office, he paused and asked, "So what does Laura want with these reports? Do you have any idea? We're going to run through this data, but how many of them should we group together? What kind of format does she want?" Those are all the right questions...almost.

Asking the right questions leads us to answering the right questions. At times, even though we ask the right questions, we cannot immediately get an answer. This is often the case when guessing what our manager wants us to do.

When we can't immediately get an answer to our broader questions, such as "What format does my manager want this in?", we need to ask (and generate an answer to) more specific questions that can be answered with a simple Yes or No. "Does my manager want this data compiled and presented according to each individual test? Does my manager want an overall representation covering the entire last year?"

Open questions lead to detailed questions. Detailed questions that can be succinctly answered with definition are where decisions are made.

Yes or No questions are the questions that we need to generate and answer when guessing what to put together for our managers. We do our best to put ourselves in their shoes, and pass judgment on the idea we have generated to propose to them. "Does my manager want these charts separated by thermal zones?" Yes or No. "Does my manager want the control sensors all grouped together?" Yes or No. "Does my manager want me to spend an extra three hours to put these charts onto a PowerPoint slide so she can use it to present the data to other engineers?" Yes or No. "How about five extra hours so it really looks fancy?" Yes or No.

Narrow your broad questions into smaller questions that can be answered in Yes or No. In doing so, you will find you can adequately guess the answers to most of the questions you were going to ask your manager. Then you can take the important questions to your manager. If you guessed wrong on other questions, it will become apparent when you talk with your manager.

Anticipating Needs

Even after you've begun finding your place with the team, sometimes you can be left without direction. When this happens, it's time to get creative again, and start asking questions.

I had been working about a month and was momentarily between assignments, when a bank of heaters failed in our primary test chamber. The failure occurred while a payload was in test, so everybody who knew anything about the system was working to sort out the problem. That immediate concern took precedence over everything else. I was too inexperienced with that system to be of any value, so I was left entirely alone for the next few days...with nothing to do.

Nobody likes sitting idle at work (that is, no engineer likes sitting idle at work), so I started looking around to see if I could find something interesting and useful. I wandered back to another test chamber we were upgrading. I hadn't been doing any work back at this chamber, but a week earlier they had mentioned something about new liquid nitrogen lines they were putting in, and suggested needing a solid model of the system. They had a big board near the chamber with the plumbing schematic for the nitrogen lines pinned up on it. That looked interesting, so I started trying to decipher it.

I'd never done anything with plumbing schematics before, so the symbols were entirely foreign to me. Making sense of the schematic was just like trying to figure out thermodynamics homework after missing the lecture because I was on a climbing trip. I just started making logical connections between notations and guessing how it probably functioned.

In addition to my own guessing, we had a few technicians who were working on the chamber. They were familiar with the symbols of the schematic, but even more valuable, they were familiar with how nitrogen systems actually worked. Liquid nitrogen is very cold, which means it doesn't take a lot of time for it to flash into a gas. When it flashes into a gas, it has a 1/1000 expansion ratio, creating enormous pressures, in turn creating potential explosions all over the place. Nitrogen plumbing needs to be designed with this and many other factors taken into consideration.

From the technicians, I learned a lot about working with liquid and gaseous nitrogen. Wherever you work, you will find technicians who know a lot about their job. They may not have a degree, but they have engineering smarts. Take every opportunity you can to learn from them about the practical side of your projects.

From the schematic, I learned what kind of functional system they were planning to implement in this chamber. I saw why they wanted a solid model, so I started creating one. I emailed my manager to inform her I was doing this, but they were so involved with the heater failure, all I got in reply was, "Great. That'll be fine."

Creating this model turned out to be the best thing I could have done. When they resolved the heater failure, they dropped in to see what I was doing, and let me continue forward with it. I ended up completing the model at just the right time for the technicians to begin installation on the chamber.

Not everything you guess at will turn out to be successful. Sometimes you may guess wrong, and the work will be abandoned, but usually something good will come from it. When you take the initiative to find what needs to be done, it helps your manager with her job. Just stay in communication with your manager, and you should be able to hold to a useful track.

The Challenge of Management

There are some important things to recognize in the interaction between you and your manager. Both of you share a joint role as employees of the company. You have been assigned to your manager for doing what needs to be done. Your manager has been given responsibility for directing you, for knowing what needs to be done (at least that is the way the organizational charts define it). In application, the managerial role is much more a shared role between the team and the manager. We'll try to understand how this works.

No manager is all knowing.

Surprising, isn't it?

As team members, we would like our managers to be all knowing. As managers, they would like to be all knowing. But the reality is that they're not. All their knowledge of the systems, status, and operation of their responsibility must come through their personal observation and investigation, or come second hand through their team members.

When I started, Lucy was my manager. Three months later, Lucy was transferred to manage a different project, and Laura was brought in from another area to take over management of our team. In this case, I got lucky...very lucky. I couldn't have asked for a better manager to start out with.

Laura didn't have any previous exposure to our team or facility, but it was expected that she would quickly come up to speed and run this operation. She had previously done a good measure of research and developed a large amount of experience that was relevant to vacuum chambers, but running a test facility was new to her. She had to rely on the expertise of the team to keep the place running initially, and also to train her in all the details of the facility. She immediately had to make decisions regarding scheduling, upgrading, maintenance, and everything else that goes along with a facility. She applied herself, appropriately scheduled team meetings, learned everything she could, and very rapidly came up to speed on everything. It was as seamless a transition as could be asked for.

This was a living example of the shared managerial role in perfect application. The manager is the person on the organizational chart, but she's more than that. She's the person who coordinates the team, makes decisions regarding the team collectively, represents the team to upper management, and so forth. The team is there carrying out their own roles, working under their manager, but managing themselves along with the manager responsible for them. The team members individually know what needs to be done, and they do it. Each individual has their sphere of responsibility, and they apply their own judgment to it.

Your manager isn't there to instruct engineers on each and every task that needs to be done. Your manager isn't there to judge concerning the priorities of each engineer's particular job. Good managers don't want to micro-manage, and no engineer wants to be micro-managed either.

Your manager is there to coordinate the efforts of the whole team, and keep them focused on the top priorities of the business.

I saw a useful analogy during a parade on the Fourth of July. The marching bands all came down the street playing their songs, and I watched one band as the conductor tripped over himself while marching backwards and leading the band. The band members no longer had their conductor up there telling them what to play, but they kept right on playing without the least hesitation in their song.

If you've ever played in a band or orchestra, you know that it's impossible to get the whole band coordinated without a conductor. Even though each individual performs his own part, the conductor is an indispensable part to the overall performance.

The dynamics are similar with a team under their manager, only engineering involves a design and operation process instead of a written script. Both the manager and the individual team members are required to make everything happen.

Let your manager know what you're doing, and let your team know what you're doing. Many decisions are made collectively by the team. Each member suggests what he thinks is best. Your manager is there to orchestrate the collected efforts of the team. You are part of the team now. Your manager and the team are counting on you to make suggestions according to what you think is best.

Individually, we hit our occasional homeruns at work. For managers to hit a homerun, it's more like something on the order of a grand slam. They have to have all their team lined up properly, and then when they do things right, the whole operation goes through smoothly, and they're excited about the success. Following is an email from my manager Laura after one of her grand slams:

> At this afternoon's TRR, Ray Russell complimented the Facilities Team for the formidable work performed after the contamination event, and most notably for continuing taking the collimator calibration measurements even after the facility had been vindicated. He also mentioned he will make sure the Air Force understands the quality and difficulty of work perform by our team.
>
> Personally I want to thank all of you. You make my work a lot easier.

Laura didn't arbitrarily instruct our team to do all the individual things we did that earned this recognition. As a team, Laura led us to discuss our top priorities and make our best judgment of what we ought to do. She does not personally know everything necessary to assure a well-run facility, but she does know how to effectively lead a team who understand their individual areas of responsibility. Everybody did a good job, and everybody within the company benefited from it.

Moving Too Fast

We're going to take a look at another experience with Mike Mamanikis, our platinum grade engineer. When he came to our team, he already had exposure to various electro-mechanical systems, and a few programming languages. Even more important than the experience, he took an interest in making computers and hardware communicate with each other.

Mike's confidence, which made him so effective, did reach a little too far in one instance and caused a minor problem that had to be undone.

When you become part of a team working on a large project, there are places where you can move in quickly and get things done, and then there are places that require a little more prudence before changing things. It is always a good idea to investigate how your actions might wrench a project before you move on them. Learning never harms anything. Analysis never harms anything. Proposed solutions never harm anything. However, when you *implement* solutions, that is when things can be harmed.

Mike had been in the control room for our primary test chamber while we were discussing some of the things we wanted to implement to keep our system current. National Instruments had released a new set of drivers, and we mentioned installing those.

The next morning, Mike didn't have anything to do immediately on hand, so he thought it would help to install the new drivers like we had mentioned. Good idea.

Mike went and installed the drivers *before* trying to find and inform us about what he was doing. Bad idea.

Seeing what needs to be done and preparing to take action on it is always a good idea. Taking action on that idea, where it affects a larger system, and without informing anybody, is usually a bad idea.

When Mike installed the new drivers, they overwrote a couple drivers from older versions that were necessary to communicate with old components in the system. He didn't know that, and the automated install routine couldn't have known that, so we suddenly lost communication with all our vacuum gauges.

We weren't in test, so this communication loss didn't hurt anything. But we had to troubleshoot the new problem, find out that the old drivers had been overwritten when Mike installed the new drivers, and then track down the old drivers and reinstall them. Nothing catastrophic under the given conditions, but in other situations it could have created real problems.

In a large system with many components and interfaces, you must know how everything affects everything else before you make significant changes. This principle holds for all kinds of designs and systems. Many factors are involved, including structural, thermal, kinematic, fluid flow, communications, electromagnetic interference, and even aesthetics. Seek to understand why things have been designed a certain way.

As a general rule, if your actions will affect a system you do not entirely understand, then be sure those who understand the system know what you're doing before you do it. Don't be afraid to *propose* a solution, or *design* a solution, but be sure your manager (and team) knows before you *implement* the solution.

Expense Decisions

The company that has hired you is in existence for the purpose of turning a profit. Projects cost money. Development and engineering cost money (lots of it). Projects must all be made to work within a financial budget and schedule. During your first year, it is unlikely that you will be making any decisions in this arena, and you probably won't gain much in-depth exposure to the financial operations. Your manager will deal with this, and your manager's manager will deal even more broadly with this. Eventually, if you climb the ladder, you will have to deal with it.

For your first year or so, you won't have this pressure, but it would be wise to pay attention whenever you hear costs and schedules being discussed. If you can gain exposure now, it will help you to make better decisions when that time comes. Pay attention to the costs and benefits of development, preventative maintenance, employee training, employee expense, overhead operations, and everything else across the board. Don't worry about making decisions, just pay attention to the decisions that are made, and try to understand why.

When you enter a corporation, you enter a world of financial dynamics that are like nothing you've ever seen before. Your company may drop half a million dollars on some facility improvements without batting an eye, and at the same time go to great lengths to reduce the cost of printing by a fraction of a penny per page. Every decision they make is aimed at making a profit, both now and in the long term.

Remember that your time is very valuable to your company. Your company is probably spending over $200,000 annually to employ you. Sound like a big number? It is. Many people fail to realize that overhead costs, employer taxes, and benefits often cost far more than their salary. For all the taxes you see taken out of your paycheck, your employer is paying roughly equivalent to that in taxes for employing you. Your employer pays for the office you're working in, power and maintenance for the building, etc.

Employing engineers is expensive. However, keep in mind that the company that has chosen to hire you is bent on making a profit. They do that by selling a product or service for more money than it costs them to produce it. If your company is still alive five years from now, it is because your work and that of your colleagues was worth more than their expense for employing you. It may cost $200,000 to employ you for a year, but the work you produce, as part of this organization, is worth that $200,000, plus a certain percentage of return. Seeing this helps you to realize the total value of your mind and your time.

When you and your manager consider buying additional products and services, keep in mind that your manager is aware of the financial goals and operations of your company. Let him make judgments on what is affordable or what is too expensive. Do your best to understand the factors involved, and how it will impact the success of the projects and the company. By seeking to understand before making a judgment, you will help your manager make sound business decisions, and learn to make sound business decisions on your own.

Wrapping up the key points of this chapter: Learn the scientific method of *Guessing What Your Manager Wants.* Ask yourself the right questions that lead to concise yes or no answers regarding what your manager wants you to do. Anticipate the needs of the team, and make suggestions about what you think will be valuable. Remember the challenges your manager is up against. Communicate with your manager, help him to be successful, and you in turn will be very successful. Offer suggestions freely, but be sure your team knows before you implement any of your ideas. The systems you're dealing with can be very complex, so be sure the right people know when you are changing things.

Chapter 10

Office Engineering—*The Hidden Workings of the Machine*

Not all of engineering is engineering. There are several non-engineering aspects that are necessary to successfully engineer a project. We have project management, business management, office management, and so on. Your skills as an engineer are important, but we cannot overlook the importance of office skills. Your ability to work within an office will have a resounding impact on your success as an engineer, so we're going to spend some time looking at the aspect we'll call *Office Engineering.*

Live Communication

Several people have pointed out that the single most important factor that will influence an individual's success in his or her career is oral communication. We will refer to what is called *oral communication* as *live communication.* Following in importance, we have written communication, and then technical ability. These latter two skills are indispensable, but live communication is by far the most important:

1) Live communication
2) Written communication
3) Technical ability

Let's take a look at what *live communication* is, particularly as it concerns engineers.

Live communication is the sharing of concepts and ideas between people, in person. It is a two-way process of transferring an idea from your own mind to others, as well as transferring the ideas from their minds to your own. Live communication isn't the ability to get your own point across. Live communication is the ability to get all points across.

When we speak of *live communication* in the field of engineering, we're not talking about being dynamic in front of a group with a PowerPoint presentation. Technical ability is more important than PowerPoint ability. Nor are we speaking about precise grammar and a vocabulary that would make your English teacher proud. These are examples of *oral communication.*

Oral communication implies communication with your mouth. You can't hear with your mouth. Since a mouth can only be used for speaking, and not for listening, it only allows for one-way communication. Communication, especially within a group, requires far more listening than it does speaking. If you're communicating within a group of eight people, your spoken part of communication will probably only be one-eighth of the process. Oral communication is about how thoughts go out of your mind.

As engineers, we need skill in live communication, about real engineering. This frequently involves charts, models, equations, and other communication tools. But there must be a live individual to communicate the science and engineering behind the communication tools.

If you think back to our original definition of what makes a project *real engineering*, you will recall that the two components are first, it is a technical project, and second, it involves working with others.

The aspect of being a technical project is all about technical ability. Your ability to understand and create engineering solutions is what drives this aspect of the project. You can't get around this.

The aspect of working with others is all about communication. You can have all the genius and engineering ability in the world, but if you cannot communicate that with anybody else, then it will never go further than your own mind.

You must be able to communicate your technical thoughts to a group if they are going to use your ideas on their project. Conversely, you must be able to collect thoughts offered by a group if you are going to take advantage of what they have to offer. Having somebody in a group does nobody any good if that person is not able to communicate in both directions.

To put it in engineering terms: Technical ability is necessary, but not sufficient, for engineering.

The ability for communication is the bottleneck in all engineering associations. It is like a bridge. Everything useful for engineering in a group must go through your communication bridge before it can be useful to you or to anybody else. What kind of bridge do you have? Is it a one-lane covered bridge? Is it an eight-lane freeway? What's the speed limit? Is there a heavy loads restriction? Can it take all kinds of vehicles (technical fields)?

If you carry the analogy far enough, you see that you build your bridge to handle several fields of thought. You have a lane for mechanical design, a lane for electrical design, a lane for programming, etc. Each of these lanes must be built, and that is what we call learning. You build lanes by learning terminology, learning to recognize equations and their application, and learning concepts and principles. You increase the speed limit by gaining fluency with the terminology and familiarity with the concepts. You become conversant in the field.

Without bridges, we are all islands. The more lanes you build on your bridge, and the faster your bridge allows traffic, the more you will be able to effectively engineer larger and more complex projects. The more thoughts you are able to send in and out of your mind and the minds of others, the more effective you will be within your company.

You may work with several people who have difficult accents and weak grammar. This is an obstacle in oral communication. Learning to understand the ideas these people are communicating will significantly increase your effectiveness as an engineer. If you happen to be one of these individuals with a difficult accent and weak grammar, then one of the most valuable things you can do is to improve your oral communication. It may take some work, but oral communication is a vital part of live communication, and as such, it will be worth every improvement you make.

Fluency of communication is the critical skill we want to develop. You want to do more than communicate adequately. You want to communicate, both expressing and understanding, quickly and precisely.

Engineering Tools—Meetings and Paperwork

There are the traditional engineering tools we think about, and those are the tools that get the most attention. These are tools like CAD software for solid modeling, structural and thermal analysis programs, and, of course, we create drawings. There are mathematical modeling and analysis tools like MATLAB®, MAPLE, Excel, and others. We have programming languages like C++, Visual Basic, and LabView, just to name a few from a long list that we use to communicate with hardware and pull data from testing. Getting back to the hardware side, we have our entire machine shop full of tools, and all our manufacturing processes for both custom and high volume products.

These traditional tools are important and useful, but there is a whole class of tools that is often overlooked as a necessary evil to engineering. Meetings are tools for engineering. Design specifications are engineering tools. Procedures for testing, for assembly, and for quality assurance, these are all engineering tools. Documentation is an engineering tool. Even software as common as a basic word processor is an engineering tool.

Meetings Meetings are an engineering tool focused around communication.

Sometimes they're just status meetings, so everybody can be aware of where everybody else is in their progress. These can seem unimportant because there aren't always measurable results from the meeting. If run effectively, these can be highly valuable in many peripheral respects. At a minimum, this is a time for your manager to make sure he's on top of everything everybody is working on. Your manager isn't *all knowing*. Since he's not all knowing, he needs some kind of tool to make him *adequately knowing*.

The fun meetings are design meetings. Everyone gets wrapped up in a storm of technical debate, searching to anticipate problems and create solutions that are impervious to error. Individuals generally appreciate it when another points out a flaw in their design idea and contributes better suggestions. We get to combine the genius of several minds to generate flawless products that perform exactly as we envision.

Similar to design meetings are meetings for testing, analysis, and so forth. Engineers like to test their equipment. If it were left purely to the engineers, they would test their products to death. However, testing is expensive, so business managers need to drive engineers to decide which tests are most important. These meetings are held to consider the design constraints of the project, in perspective of the design constraints of the business. While engineers are trying to maximize performance, the business managers are trying to maximize profit.

Then there are the dull meetings: procedure reviews, process reviews, and so on. We go through these to make sure that what we've decided to do really is right. Boring as they may be, it's a bad idea to neglect these. This is your *idiot check* on your work. No engineer wants to have his name on the Monument of Stupidity. When we adequately review our work, both individually and collectively, we catch those things that we *almost* did to earn our place on that infamous monument.

We often think of meetings as formal things, but they really are about communication. Most of the design meetings engineers have are in the hall, passing from one place to another, or right there with the hardware, or in front of the computer displaying the data for you to see the first time. Engineers need to communicate all the aspects influencing the projects they're working on. Whether this communication happens formally or informally, it needs to happen.

Requirements and specifications A whole discipline of engineering has grown out of the industry need for Requirements and Specifications. As technology in every field progresses, so do our demands to integrate all these technologies into a single product.

It used to be acceptable to make a product that did one thing and did it well. No longer. Just look at your cell phone. We want our phones to take pictures, send videos, surf the internet, check email, buy and sell stocks, be an alarm clock and calendar, keep contact information for everyone we know, carry a calculator, and we need our own touch of personality and style. Oh yeah, they still need to send and receive calls (but we *must* have custom rings, and pictures for different callers).

The days of single function products are gone. The days of single field engineering left with the days of single function products. All engineering has become multidisciplined, and the discipline of engineering for dealing with that is Systems Engineering.

Systems Engineering focuses on defining the complete range of customer needs and required functionality for a product, documenting requirements, then proceeding to implement and verify the fulfillment of these requirements throughout the design and integration process. Put simply, Systems Engineering defines what your product will do, and makes sure that it does it.

When the military wants a new satellite system, they contract with an aerospace company to build one. The military branch is the customer, and the contracting company is the supplier. It takes several years to develop the satellite, as well as the ground support systems to talk to the satellite and process the data it sends back to earth. Determining exactly what this satellite will do, and what they'll be able to do with the data it sends down, is a complicated process. Then carrying those requirements and specifications all the way through the project to ensure that each subcomponent will deliver as promised becomes even more complicated. It's a formidable engineering challenge.

When the Systems Engineering is done right, and the Requirements and Specifications are successfully threaded through the entire design, then the projects are successful. When the Systems Engineering is botched on any level, it turns the whole thing into a debacle.

The Air Force contracted with several layers of aerospace companies to build a new Space Based Infrared Satellite surveillance system (SBIRS). The project started smoothly enough and ran well...until it was time for testing.

The Systems Engineering hadn't been properly executed. The Requirements hadn't been adequately defined up front, nor had they been properly threaded down as the design process became more detailed. It cost hundreds of millions of dollars and over four years in schedule delays before the project was back on track. It wasn't a failure in technology, or failure in individual systems, that caused the problems. The ground level engineers all did well with their responsibilities. The problem was rooted in the initial documents defining the customer needs.

When professors want you to create a Customer Requirements Document for your Senior Design Project, they're trying to introduce you to the larger world of engineering. These requirements are the tools that will be used in industry where the projects are larger and more complex than anything you've worked on in college. Creating such Requirements and Specifications for your Senior Design Project might feel like using a sledge hammer to tap in a nail for hanging a picture. Overkill. However, you're not creating the document because your project needs the tool. You're creating the document because you need to learn to use the tool. When you enter the field, such documents will become absolutely essential.

Procedures As an engineer in industry, there will be procedures created for nearly everything. Much of the procedure will be painfully obvious to anybody who is familiar with the product and the test setup, but the procedures are necessary. Essentially, procedures are a check list.

Designing procedures is critical, whether they be test procedures, assembly procedures, shipping procedures, or anything else that is performed. It requires an engineering mind to organize everything that needs to be performed, and to create a document that delineates each step in such a way that there is no ambiguity or confusion.

When creating a drawing for a part from a CAD model, we follow Geometric Dimensioning and Tolerance standards for representing every feature on that drawing. Creating unambiguous procedures is of equal importance as creating unambiguous drawings. If the drawings aren't correct, the part won't be built correctly. If the procedures aren't correct, the product won't be assembled, tested, or operated correctly. Everything must be defined clearly. Once everything is defined, the steps must be followed correctly.

Often, testing is very involved and very expensive, and it is easy to miss a step. Most tests on satellites run several months. It's one thing to know what you want to test when all you're interested in is a single objective five minute experiment. For several months of experiments, even the most attentive engineers need to create a procedure to ensure they cover everything.

Some testing is very short in duration, but the setup process is very lengthy. Automobile crash tests are an example of this. One engineering team got their vehicle all set up and instrumented for an impact test. They crashed the car, the test was executed perfectly, and their data came in just fine. Then when they started into the analysis of the accelerometers and strain gauges on the frame, they realized that nobody had taken the time to document which instruments were wired into which data ports. There they were with hundreds of streams of data, and it was useless because a single document hadn't been created.

Procedures for very routine tasks are also important. We had a sensor that came out of Northrop Grumman where I started working, and it had been integrated, along with components from several other companies, onto a satellite where another company was the lead contractor. Then one day I walked into work and saw a picture of that delicate satellite crashed down on its side. All the instrumentation, electronics, and everything else was shiny and pristine...crashed out on the clean room floor.

Ouch!

That day the whole aerospace industry got to look at one single accident near the end of a long sequence of meticulous design, integration, and testing. They saw $233 million in damages, just like that, faster than even the stock market can crash.

This accident wasn't caused by a large oversight; it was caused by something very simple. When they moved the spacecraft, they didn't follow the Move Procedure. That procedure included a simple step of checking that a few bolts were in place and tightened. It's a no-brainer step. Anybody can see that the bolts that hold down a satellite need to be tightened. But when you combine that single no-brainer step with a hundred other no-brainer steps, it's easy to overlook one of them; and one oversight is all it takes.

Accidents like these can happen at any company. It's not a result of bad design, bad management, or some kind of complex business process. It's the result of a few people getting into their routine, and forgetting the importance of procedures.

Documentation Technology changes, and when it's time for upgrades, you need to know how your existing system was created so that you can know how to integrate new components. We make design documents for two reasons. The first obvious reason is to build the project the first time around. The second purpose is for when it's time to make an upgrade. It's one thing to have existing hardware, but if you don't have documentation of its design, it becomes exponentially more difficult to make modifications.

We were making a new infrared radiator plate for some testing we wanted to do. At the time, I was the only engineer familiar with Pro-E, so they threw the project at me. It was just the radiating face that we wanted to replace. The rest of the system was working just fine, so we wanted to leave it intact.

When they made the original radiator plate four years earlier, they included drawings with the documentation, so we pulled out those drawings to find the bolt patterns and dimensions for the new plate. The drawings they made were for their own machine shop, and we found that there were two very important dimensions missing. I imagine that the machine shop called the original designer for these dimensions, and they were given verbally over the phone. The part was made correctly, but the dimensions never made it to the drawing. This created a problem for us.

The measurements we needed weren't dimensions that could be logically guessed. We thought of removing the original plate to take measurements, but it's in a tricky spot with delicate instruments around it. Removing the plate for two measurements, and then putting it back for use until we had the new plate, would have been a risky 20-hour ordeal.

Fortunately, one of our engineers was able to look back through some old computer files and find the original AutoCAD drawings. We pulled the dimensions from the original drawings and everything worked out, but if we hadn't been able to find those files, we'd have been up a creek. Neglecting simple things like that can become very expensive in the long run.

The moral: Make sure your documentation is accurately completed, and update everything with any changes that are made. It may seem like a tedious waste of time, especially after the product has been made, but it could save a lot of time down the road.

Document events Just like we need documentation of equipment, we also need documentation of any testing. This is a lesson you will learn on your own after analyzing the events of a few tests, but it will make your life far easier during your first year if you start off doing things right.

When you are engaged in a testing situation, fixing complicated hardware, or building hardware, it will greatly help if you thoroughly document your actions and the things you notice throughout the process. If by any chance you need to remember what happened at what time, or what wires went to which connectors, a physical document will do wonders for confirming your memory.

The following exerpt is from an email a manager distributed to everybody involved with a test shortly after some anomalous temperature activity had been observed. All those involved had paid attention to what was happening at the time, but when they tried to piece the situation together after the event (looking for the root cause of the anomaly), there was no record of many things that were done. Without having a record of many of the causes controlled in the system, the recorded effects were of little value:

> This is a plea to all of you to improve your documentation in the facility log books at WC-2. It is critical that each shift accurately documents ALL changes made to the facility settings, or even observations made during your shift, in the facility log books so that future shifts are aware of what has gone on during previous shifts. I've emphasized to the test conductors and shift bosses the importance of communicating what's going on. It's important that you're all aware of what operations are occurring during any of the shifts you are assigned back at WC-2. Adjustments to settings, no matter how minor, must be in the log book! We've had some temperature "excursions" over the last couple of days go noticed, but not documented

> and causing concern to subsequent shifts. We've also had adjustments made to valve settings, and temperature settings that have not been appropriately documented as well.
>
> Thanks in advance for your support.

It is likely that your company already has processes and procedures in place for keeping such documentation. Be sure to use these systems.

When you're new, you may be given the freedom to run several tests on your own, independent of the procedures traditionally required by your company. Keeping your own log book during these experiments may not be required, but it will be to your benefit.

When you were a little kid, did you ever take apart your dad's lawn mower, or your mom's blender, and then find you couldn't put it back together? If you've made that mistake once, try not to repeat it when you start as an engineer in industry. Documentation is how you save yourself from repeating this mistake.

When you're new, you may be given the assignment to troubleshoot or upgrade equipment that nobody else is familiar with. Since nobody else is familiar with the equipment, there won't be anybody to answer your questions if you are uncertain about how to put things back together. Keeping your own log book of the process may save you the embarrassment of creating a larger problem than the one you started fixing.

Office Engineers

So who does all this *Office Engineering*? Engineers do it. Obviously, it's not something that can be handed off to an English graduate. It takes more than good grammar to run the paperwork in an engineering department. Shakespeare may have had some excellent insight into the interactions between people, but we need people who have studied the interactions between mechanisms.

A lot of paperwork and meetings are necessary to run large projects, and people with a tolerance for this *Office Engineering* find their way into these positions.

Stephanie started just like the rest of us. She was hired as an engineer, could have gone any direction, and ended up taking the meetings and paperwork route.

It was nearly a year after she started that I was asking her how she can tolerate all the meetings she spends all day preparing for and attending.

Her laughing reply: "I don't want to be an Engineer!"

Stephanie's manager didn't hire her to be an *Office Engineer*; she hired her to be an engineer. Stephanie grew into the position, just like everyone else grows into various positions. She is fine with doing paperwork and attending lengthy meetings. She does a good job with it.

At the start, Stephanie attended the regular meetings and didn't complain when she was given paperwork to update. Being reliable with an assignment is a sure way to be asked to perform the job again. She wasn't really busy at the start, but as she learned what was going on with the projects, and her managers learned they could rely on her to do the job right, they transferred increasing responsibility to her.

Over the course of several months, Stephanie merged herself into the precise position that another engineer had been performing. This other engineer had begun trying to move to a project she was more interested in, and having Stephanie there to incrementally pick up her responsibilities for these meetings made the transition easy for her. Nobody made a decision to move Stephanie into this position; it just happened as each person did what she was most interested in doing.

All the meetings that Stephanie spends her time in are associated with integrating and testing our satellites. Her time is spent creating and reviewing test procedures and schedules. She does paperwork. Not *busywork*, but *paperwork*. It is also important to recognize that this paperwork isn't something that we can hire a business student to perform. This work requires an engineering mind. It requires somebody who understands the systems we're working with.

Through only a couple years in this position, Stephanie gained exposure to the technicalities and testing of half a dozen satellite systems in various phases of development. She has become familiar with all satellite testing chambers and systems. She deals with the payload itself, the data it sends for both the mission and its own health and status, as well as the testing chambers that create the testing environment.

When any anomalies occur, Stephanie is right there. When we're rearranging schedules or testing sequences, she's involved. And when we're putting together the procedures of what to test, when to test, and how to test it, she's right there in the details. It's her engineering mind that makes her so effective with this work.

Technical writing It's sometimes funny how true to form engineers are. In general, engineers don't like to write. They like to write equations, or draw graphs, diagrams, and flow charts, but they don't like to use words unless absolutely necessary. When put to the task, many engineers will agree that writing for engineering is difficult.

Understanding can be communicated with incredible clarity and accuracy using an equation or a drawing. To try to communicate the same information using only words is an exceedingly laborious task, not to mention slow, cumbersome, and fearfully susceptible to ambiguity and misunderstanding. However, even with the clarity of drawings and mathematics, some things remain that can only be communicated using words.

Ask any salesman of technical equipment, and he or she will tell you the trick to selling technology is being able to "translate Geek into English." Translating Geek (or rather, science and engineering) into English truly is an art. Not many people develop the talent to understand science and engineering. Among those who develop this talent, we need those who can further develop the talent to translate it into English.

Just like we need engineers who can produce concise drawings accurately and rapidly, so do we need talented engineers who can produce documentation and reports accurately and rapidly. When it comes down to it, writing as an engineer is nothing more than communicating the same ideas and principles, but using the medium of words instead of equations or drawings.

During your whole college career, your professors beat and batter you until you figure out how to understand engineering using math, a language only other engineers understand. Your professors train you to think and communicate in the languages of science and engineering.

Then as your last year approaches, your professors start switching their emphasis and require that you begin writing what you've been learning, using English. We finally become accustomed to this new language of engineering, and then we're asked to transition back to our verbal language. Your professors start coercing you into this transition because they know the demands of industry, and they understand that in order to survive, you must be capable of communicating in both forms, calculative and written.

Open up any one of your textbooks, and open it to any page. Chances are there are one to four equations on that page, and four to eight paragraphs explaining those equations. This isn't a coincidence, and it isn't because the author felt like rambling for a while. It is because each of those equations requires four to eight paragraphs to explain what it means.

When you begin writing documentation and reports, the ratio of substance to words will remain about the same as it is in your textbooks. This may feel terribly inefficient at first, but it is absolutely essential for somebody to understand what you are reporting on. You have learned to understand clear and concise writing from your textbooks; now you must learn to produce such writing.

You've earned your degree and learned to speak Geek. This served you well through college. Now you're in industry, and you must learn to translate Geek into English.

The most important thing to keep in mind when you are writing is *who* you are writing to. *Who* is going to read this? *Who* is going to pick up this documentation or report and decipher what is written?

Who is this person?

Is he a technically oriented engineer?

Is he a manager who has been removed from the technicalities of the project?

Is he an executive who wants a concise summary?

Is he a person 5, 10, or 15 years from now who will be making modifications to what you have designed?

Answer *who* will use this document, and keep that person in mind while writing. Knowing who will use the document will guide you to how much detail to include, how much you can write in Geek, and how much you should translate into English.

Writing for engineering can be difficult, but that doesn't make it any less important. Writing is just one more skill that you need to stack onto your degree. You learned differential equations. You learned thermodynamics. You are also capable of learning writing, so stop whining and do it.

Company documentation and reports Most large corporations will already have a system in place or format they want you to follow for your reports and documentation. Even though these systems have been developed by engineers and managers with years of experience in the field, they still will each have their advantages and disadvantages. They probably won't be perfectly suited to you, but don't let this bother you. The best thing you can do is to find a way to go along with the system.

For all the complaints your more seasoned coworkers may have, the systems really have been designed to do a pretty good job. In your university, you learned to conform to the content and format the professor wanted. In industry, you'll use that same learned ability. Conform to the content and format your company wants, and life will be easier for yourself and everybody else.

Personal Finances

When you start with your company, for the first time in four years you will be earning a substantial paycheck. Instead of living on a negative cash flow with your student loans, or scraping by with your part-time job, you will be making more money than you previously knew what to do with.

It feels good to be earning an income, but if not managed properly, it can get away from you. When you feel like you have a lot of money, it's easy to stumble into a position where you feel the weight of debt more heavily than you ever felt it before. This is something you can avoid if you keep your head screwed on financially.

Everybody has a friend (or maybe you were the friend) who starts his freshman year at college, goes crazy with the freedom he's never had before, and spends the rest of his academic career making up for the miserable GPA he earned those first two semesters. It's a poor way to start college, but it's not as bad as starting your career that way. If you make poor financial decisions at the start of your career, you can get yourself into a hole, and it can take a long time to climb out.

In addition to enjoying your job, you want an income sufficient for you to live comfortably. Let's talk about what it means to *live comfortably*.

You can have a new car with all the bells and whistles, and it will feel pretty comfortable riding down the road. But how comfortable will it be when you're trying to figure out how you're going to make the payment on it next month? How comfortable will it be if you have three credit cards, all carrying balances with high interest rates? How comfortable will it be to bypass the next trip your friends are taking because you can't afford the flight, and your car, and your credit cards? It's easy to acquire one comfort, and bring along with it seven other discomforts.

"Interest never sleeps, and it never stops working."

This piece of wisdom from personal finance gurus identifies the dual power of interest. Interest has the power to hold you in debt, or to lift you to wealth. It's all a matter of whether interest is working for you, or against you.

Financial stress doesn't leave you alone when you're at work. During the eight hours you're at work each day, your car sits in the parking lot, your big screen sits in your apartment, and Hawaii sits in the middle of the ocean. If you owe money for these things, the debt sits with you there in your cubicle. Financial stress will negatively affect your performance at work, which will in turn negatively affect your future income.

Financial stress is one of the heaviest discomforts people can pull onto themselves. Escaping financial stress doesn't come from making more money, but from being discriminate with the money you have.

Here's the advice:

Engineer your finances!!!

One of the most effective ways to keep yourself in a position to live comfortably is to model your finances, create realistic financial goals, and stick to them. It's not difficult to build reliable financial models. Even people who are mathematically challenged are still smart enough to use a spreadsheet to model cash flow.

Creating a spreadsheet accounting for how much money you have coming in, and your expenses going out, is simple. You can set up the model in a recurring fashion on a weekly or monthly basis. Your income section will be a simple one or two cells. Your expense section will include things like rent, student loans, car payment, car insurance, food, etc. Several software packages are available that will greatly help you track and forecast your finances. A few bucks on one of these packages would be one of the best investments you could make.

Financial goals are things like paying down existing debt at a target rate, building investments at a rate, holding certain expenses to a chosen limit, etc. Having *X* dollars invested by *X* date is not a realistic goal. However, investing *X* dollars on every *X* day of each month is a realistic goal.

When you create your financial goals, list them in an order of priority. There will be anomalous events that force you to break some part of your goals, so prioritizing them is how you bridge these events without breaking your entire plan. If you have existing credit card debt, paying this down would be your number one priority.

A general rule about toys and vacations is if you can pay cash for something, you probably can afford it. If you have to finance it, think twice, and then a third time, before committing yourself to it.

This doesn't mean you have to be tight about your money. Actually, it gives you the freedom to be quite the opposite. If you get yourself laden with debt, then every penny becomes critical to you. If you keep your debt level low enough that you can easily pay your obligations and still have an excess slush fund, then it takes more than a few bucks to push you into the red.

In school, you learned the principle of Conservation of Energy. Energy in equals Energy out. With finances, we apply the principle of Conservation of Cash. Cash in equals Cash out. Buying toys faster won't allow you to buy any more toys than you could buy otherwise. When interest gets involved, buying toys faster decreases your capacity for acquiring more toys.

If you will take the time to engineer your finances, you will be able to have all the toys you can afford, and you will live far more comfortably than you would be if living under debt.

Engineering a project requires a whole office infrastructure to be successful. *Live communication*, the ability to get all points across among people, is the most valuable tool you have for office engineering. Meetings, paperwork, procedures, and documentation are all necessary tools in the engineering process. Technical writing is a skill every engineer can learn, and needs to learn. Ask who your audience is, and write to communicate accordingly. And finally, engineer your personal finances so your mind can be free while at work, enhancing your performance, and thereby your overall income.

CHAPTER 11

"Houston, We Have a Problem."—*Changing Programs*

Most of the time, engineers are able to find their role in a team without unusual difficulty. You might experience a few bumps, bends, and false starts, but within a couple months, everything works out. However, sometimes the match doesn't quite materialize. The mismatch can be from lack of projects on a team, lack of challenge, or other unforeseen scenarios. When this is the case, it may be time to change the program you're working on.

Help from Management

Nate Hopkins joined an aerospace defense company at the beginning of the summer as a mechanical engineering intern. He was only there for the summer, but his experience could just as easily have fallen to a newly hired full-time engineer.

Nate had previous experience using Pro-Engineer, so they initially placed him in the machine shop and made him responsible for designing a fixture for tensile test specimens. Nothing glamorous, but it was useful and required some design. They gave him the specifications, the samples that they would break in the fixture, and let him go at it. This was productive and interesting work for the two weeks that it lasted.

Once he finished that project, his manager didn't have any other work to give him. Every once in a while he was given a part to program a CNC path, but nothing intellectually demanding. He made good use of his time, learning Pro-Mechanica and more tricks in Pro-E, but he was itching to do some real engineering.

Nate talked to his manager, who eventually resorted to giving him secretarial work: organizing papers, filing things, etc. It was work that needed to be done, but certainly not engineering. Nate and his manager contacted a couple other managers to see if they had work Nate could pick up, but to no avail. Nate had been placed back in the machine shop, and had few meetings to attend, limiting his exposure to other managers and other projects.

In most situations you can stay positive and offer to help out other teams as you see their needs. They'll be happy about your interest and desire to work, and as you deliver quality results, they'll hand off more work to you. In such cases, you can make a seamless transition to a project where there is work to be done. However, this is not always the case.

Nate had been there two months, so his manager suggested he speak with Mary in Human Resources (HR). The HR staff is there to help out with situations like these, and they have all the connections to be very effective. Mary suggested a manager for Nate to contact regarding the type of work he was interested in. Nate went and talked with this manager about the work he had been doing, his performance on that work, and the type of experience he was hoping to gain.

When Nate walked into work a week later, there was a message waiting from the program manager in charge of Advanced Systems for Smart Munitions. Nate joined the group, and they immediately teamed him with another engineer to design and build a custom test setup for a developmental smart weapon sensor assembly. They designed and built the machine to spin the sensor at 30,000 rpm, the fixture to hold the sensor, and the data acquisition configuration. Nate's experience with Pro-E, his time in the machine shop, and his classes on data acquisition and programming came in very useful.

When they ran their test, everything went well. They collected the data, and sent them to several new computer science graduates to work their magic manipulating the data for developing the guidance algorithms.

While working on this, Nate was also given a couple other projects dealing with this same development system. He had proven his integrated mechanical aptitude, so he was asked to design a mechanism for transport and loading, and attaching the electronics to send the launch signal. Nate worked with the senior engineers to be sure it would function with the complete system, and they adopted the design he created. Just before he left, they had him develop some concept ideas for another component, which he had to leave for another engineer to finish.

It is significant to point out that all this was done in his last two months with the company. Nate took the initiative to get himself assigned to a project where his skills were valuable, and it ended up being better for everybody involved. Nate had a blast with the work, the company gained more value from his time, and his old manager didn't have to put up with the headache of finding secretarial work for his engineer intern.

It's never too late to become productive, and if you're in a position where the company isn't benefiting from your presence, chances are that you can talk to the right people and find a place where you are more valuable. Before contacting other managers about projects, be sure you have spoken with your own manager and made every effort to work through his direction. Very often, mismatches can be taken care of right there at the ground level. Nate didn't cross over any lines without his manager previously being aware. It is in everybody's best interest for you to find a position where your individual abilities are needed.

Patience Is a Virtue

Ambition works well for getting things accomplished, but sometimes it is best to simply be patient. When you step into your team you've been hired to work with, there may not be anything exciting and challenging that is readily apparent to you. Even though the team feels overworked, and they need another member, they may not know what to do with you right at the start. You may feel inclined to start looking at other teams where responsibility is more readily available.

This is the time to be patient. Not *apathetic*, but *patient*. Apathy is sitting and waiting. Patience is working and waiting. If you are being patient, all the time you are waiting, you are keeping your eyes and mind open for opportunities to get what you want, and being diligent with what you have until the opportunity comes by.

We all have good days at work, and we also have bad days. When we're on a bad day, sometimes things can happen that magnify our current state of discontent. About eight months after graduation, the following email showed up in my box from my good friend Katie Timothy:

> So here is what not to do. Yesterday I told my stand-in boss that I don't have any work. Since my real boss is away on vacation for the week, I am kind of leaderless. Anyway, today my stand-in boss gave me the task of updating the weekly drawing tree. This doesn't sound like any fun at all. It will probably take me 15 minutes once I really sit down and do it, and I have nothing else to do for the rest of the day. I really need to get on a new program here at my company. This one is slowly draining my life force away.

Few people speak their minds so point blank. In addition to being a brilliant engineer, Katie is one of the most entertaining writers you'll ever read.

Katie was hired by an aerospace company in Colorado, and her lightning skill on a few CAD programs landed her with the responsibility for modeling components of several satellites. She made changes and modeled things faster than anybody else could even sketch them, so they used her for what she was good at. However, after you get yourself situated with a team, and you have found a role you're productive with, that doesn't necessarily mean you're going to be enjoying it.

Initially, she enjoyed her role because she was in demand, doing important work, and visibly productive. But it wasn't challenging. She wanted to be more involved with design and analysis.

Katie was patient, but wasn't apathetic. She let her managers know what she was interested in. It took a little while for the response to come through, but when the opportunity came, she was tapped for more demanding responsibilities.

Her familiarity with the spacecraft flight mechanisms and the test hardware, developed through her modeling work, had ideally positioned her to take on greater responsibilities for these projects. Katie became the primary designer for several components and tooling on two separate projects. With greater intellectual demands, and still with her lightning ability to use the design software for both modeling and analysis, she was able to become far more valuable to her company in the very program that was previously leaving her bored. She let her managers know she wanted something more challenging, and they responded as soon as they could. Katie never sent anymore emails of boredom after that.

If you perform well at whatever task you are given, it will open doors in the future for obtaining the positions you really are interested in. Sometimes those doors aren't immediately available, but outstanding performance makes you ready for the moment they become available. Sometimes, being patient is the fastest way to get what you want.

When the Time for Patience Has Expired

Patience is only a virtue if there is no faster way to get what you want.

Sometimes, after being patient while working diligently and letting your managers know that you are interested in something more challenging, it becomes apparent that you're not going to get that challenge at your current job. When this is the case, it's time to take action. Waiting around any longer ceases to be patience, and becomes apathy.

Matt Keller started with a major aerospace company at the end of the summer. He was hired into the Mechanical Engineering department, and after a couple months they didn't have an immediate need for him, so they sent him over to the Payload Integration and Test team. He worked on several small projects and was as bright an engineer as any. In spite of his good work, circumstance landed him in a place where his manager was unaware of what he was doing, and unable to place him on a larger project suited to his interest.

Over a year passed, and Matt was still working in a different department than he'd been hired into, was unpredictably split between two managers, and had no sign of anything changing in the future. It was time for him to polish up his resume and start searching.

The whole job search and transition was through in two months. In addition to positions he applied to over the internet, he contacted a manager at another facility who had offered him a position a year earlier. He explained the situation, and this manager was eager to bring him onto his team.

Making the transition went relatively easy, and he took a position working on a high energy laser countermeasure system. This is a chemical laser that automatically senses rockets or artillery, focuses its laser on them, and tracks them until they're blasted out of the sky. It's sort of like a laser performing the job of a Patriot missile. High tech, high precision, and like all systems of that nature, it's fun. Matt made the transition, and immediately began enjoying his new job. He was put on some challenging projects, and he fit right in.

It is important to recognize that Matt's manager was not surprised when he made this transition. His manager was pleased with the work Matt had done, and he had been trying to find a more involved project for him, but was unable to due to the circumstances. Matt had been both patient and diligent, and nobody's feathers were ruffled when he made the transition.

Cautions About Changing

In reference to changing programs, you want to be careful of leaving a job early. Such a transfer could be a yellow flag on an individual. Unless there is some good reason for the transfer, employers are initially leery about such moves. But if the situation isn't what you want, you may be better off cutting your losses early. Every situation needs to be considered in its own context. For Matt, it was best to leave. For Katie, it was best to stay. You will get to make your own judgments.

When you inform your current employer that you are leaving for another job, sometimes you will be given a counteroffer. The counteroffer will usually entail an increased salary, but it could also include a change of responsibilities, opportunities, etc. If the counteroffer is enough to make you reconsider, ask yourself a few questions before swaying one way or the other. Why did you want the new job? Under the new conditions, which position will you enjoy most? How realistic are your expectations for the new job, or the promised changes with your current job? Are you being given the counteroffer because your current employer values your work, or because your current employer is desperate to keep the position filled? Lean toward the position that you will enjoy, and where you will perform best.

Above all, don't waffle on your decision! If you decide to accept your current company's counteroffer, stick with it. You can discuss your counter offer with the new company, but returning to your current company with a counter-counteroffer will seed ill will. Also stay away from continual job shopping. When you commit to a company, stick with your decision for the appropriate timeframe.

There's one more thing to keep in mind when changing programs, or changing jobs after a few years. Most companies require a two-week notice before leaving. Two weeks is all they ask, and two weeks is probably the earliest notice they want you to give them.

Engineers are courteous people. We work well with our teams and managers, so we don't want to leave them up a creek when switching jobs. It seems like a considerate gesture to let your manager know as soon as possible that you will be leaving at some future time. This seems like a good idea, but in practice, it may not be all that hot.

People are what drive engineering projects. A person can be a significant part of a project to replace. But, if a relatively new engineer must be replaced, two weeks is probably sufficient time to get the critical responsibilities transferred. Less time isn't enough time, and more time throws a wrench into the works longer than necessary. As soon as your manager knows you're leaving, he can't give you new or continuing responsibilities, and you become dead weight on the team. Situations vary, so be sure to think of the team dynamics before making any announcements.

It is also a bad idea to tell your manager you will be leaving if you aren't absolutely committed to walking out the door on that date two weeks away. Be sure you have your next job lined up before you make any comments about leaving. Sending the message that you will be leaving at some unpredictable time does nothing but stress your manager. You may think you're doing him a favor, but the result just makes his life more complicated.

A couple things you can do to prepare for leaving, which won't interfere with your team, are verifying your unique knowledge is documented, and organizing your responsibilities for a clean handoff. These are things that you would do anyway, and if you don't leave, they help with your job as you continue where you are.

If the worst case scenario arises and your program just isn't working out, your manager is your first line of assistance for getting repositioned. Often, working diligently while patiently waiting for your manager to find the right opportunity is the best thing you can do. Let your manager know you're interested in more challenging work, perform well with the work you have, and opportunities will arise. If it becomes apparent that leaving is your best choice, be sure to make the transition appropriately, considering all necessary cautions.

CHAPTER 12

Getting Smarter—*Continuing Education*

One of the questions facing every undergraduate in engineering is that of an advanced degree. Do you get a Master's or Ph.D.? Do you get a job first, or go straight through with school? Do you earn it part-time or go back to school full-time? Do you get a Master's in Engineering or Management?

There is no *one size fits all* for education. Everything depends on you individually, what you enjoy, and what your goals are. Bill Gates dropped out of Harvard to start up Microsoft. Jack Welch (CEO of GE for 20 years) decided to earn an engineering Ph.D. simply because he thought "Dr. Welch" sounded better than "Mr. Welch." Many M.B.A. graduates argue that individuals with a Ph.D. are overeducated for business purposes, but most aerospace executives have engineering Ph.D.s.

The only constant concerning education is that it's a stepping stone. What you do after you have your degree determines what happens with your life. You will learn much while earning your degree, but you will continue learning across several fields throughout your life. Regardless of what field your degree comes from, and what level your degree is, you will learn many highly important things throughout your career.

Master's Credentials Are Minimum

The one rule that engineers and managers in mature industries agree upon is that a Master's degree is worthwhile. It doesn't always matter if the degree is technical or managerial, but the credentials of a Master's degree are essential to certain opportunities.

Phil Cojanis earned his undergraduate degree in Mechanical Engineering and has worked as a systems engineer for several companies on many projects. He has continuously been involved in training courses to learn specific skills and earn certifications in various technical and managerial fields. In addition to his engineering career, he is one of the prime movers of AIAA in Los Angeles, is actively involved in public policy regarding aerospace, and has made a difference in the industry through his efforts.

Phil has accomplished many things and learned many skills, but he never thought it was necessary to dedicate his learning to earn a Master's degree. He always felt that the ability to perform in any given situation was more important than earning credentials that weren't directly related to his immediate projects. He is right about ability being more important than credentials for engineering a project, but looking back, he regrets not having earned a Master's degree.

"It's only a piece of paper, but it makes a difference when upper management decides who gets what responsibilities."

It's easier for a manager to justify promoting an engineer with an advanced degree, than to justify promoting an equivalently performing engineer with only an undergraduate degree.

Paths Available

If you choose to pursue a graduate degree, several options are available. You can go straight through from your undergraduate degree, pursue your degree part-time while working as an engineer, or return to school full-time when appropriate. Each alternative fits different personalities and career objectives, so we'll look at a few people who have taken each of these paths.

Straight Through

James Ayers decided to finish his Master's degree before entering industry. He started with Rocky Mountain Composites, a smaller company, after finishing his Master's with a thesis involving composites. His plan had always been to continue to law school, and he had been accepted to several, but he decided to take a couple years in engineering before he left the field permanently.

The company that hired him had decided they wanted to enter the small commercial aircraft business. They had the composites expertise, stole executives from Leer and Beechcraft, recruited some seasoned engineers from Burt Rutan and other industry leaders, and started into it.

James entered the scene and hit the ground running. He didn't get any orientation or other niceties the larger corporations all start with. They threw him directly onto his own project, giving him full responsibility for success.

Before the first day was done, James had already started working on the landing gear for their new aircraft. The leading engineers got him going with several cartoons, sketches, and concepts about how they wanted the landing gear to work. The rest was up to him. He started by creating a kinematic model, grew that into a CAD model, and began running FEA (finite element analysis) to develop and refine the design.

From there, he started into generating more detailed versions of the part in composite form, and running these through FEA for composite properties. The performance of composites changes according to the way that the particular parts are built. When you design a composite part, you don't just design the shape; you design the way that shape is created. It's the fabrication process in composites that determines the strength, stiffness, and other properties of the part. You design which types of fibers to wind, the bonding matrix to hold the fibers together, the sequence, angles, and layers for winding, and so forth. Composites can give you incredible performance for your application, but the mind of an engineer is required to put that performance into the material.

As the design progressed, James modeled and ran the analysis for every component in the landing gear: the structure, the bearings, and all the way down to the nuts and bolts. James did a lot of work on his own for this landing gear, but he didn't work in solitary confinement. He was frequently consulting with the seasoned engineers, machinists, and technicians on what worked well and what didn't.

In addition to the landing gear, James was given responsibility for design and analysis on several other components, including the wings, engine pylons, winglets, and so on. He continued developing his skill and experience, and the team continued mixing him into engineering the aircraft.

Having the expertise from his Master's degree, James immediately joined an excellent design team right out of school, and progressed to play an increasingly important role on that team. He earned the opportunity to continue focusing on his career, learning what is important to his projects, with no concern about credentials in the future.

This experience is somewhat characteristic of working for a smaller company; they require more responsibility from their employees. Engineers that have been around the industry will often say that working at a small company is three times the work, and ten times the fun. At the same time, larger corporations have the infrastructure to handle projects that are far more complex and expansive. The opportunities for responsibility and technical challenge arise in a different form.

Companies of different scales also have different drawbacks. Smaller companies are not so predictable, and are more susceptible to the tides of the economy and demand in their specific market. Large corporations can feel slow and bulky, sometimes leaving engineers to feel like just another cog in the giant machine. Everyone has different strengths and characteristics, so finding the environment that is right for you is important to enjoying your long-term career.

Overall, James speaks highly of his experience as an engineer: "I like engineering. I didn't think I would, but I really like it. Designing, inventing things, doing the analysis to make sure what I design will work... I like engineering." He enjoyed it so much, he even discarded his plans for law school in favor of an engineering Ph.D.

Balancing Work and School

Ryan Woolley was coming up on graduation, somewhat burned out on school, looking forward to working, and not exactly sure where he wanted to focus for a Master's degree. He knew he wanted to get a graduate degree, but graduation wasn't the right time.

Ryan was hired by a major aerospace corporation, and learned that they would pay for whatever part-time classes he wanted to take for a graduate degree. These days, just about every major company in the U.S. will pay for part-time graduate work. Confirm this before you go to work for a company, but chances are if they're recruiting at your career fair, they'll pay for your next degree.

It isn't easy earning your degree part-time. You are responsible for learning and working at your job, and then you stack evening classes on top of that. Many universities offer classes in which you can participate through a satellite location, but you still have to learn the material and do the work.

Having gained some exposure in the field, Ryan was able to choose an area he was genuinely interested in pursuing. He chose a program in astronautics, kept up with his classes, and graduated two years later with zero debt.

In addition to the degree, he had the experience of working, which gave him a valuable perspective on the courses he was taking. He did the same work as every other student, but he was able to learn it in a context of applicability. This is one of the greatest advantages of gaining experience in industry during school. It increases the value of your learning.

Ryan was performing well and adequately enjoyed his current job. Now having gained a clear vision of what he wanted to do with his career, he knew what kind of job was more suited to his long-term goals. It's unlikely that he would have gained such a clear vision of the work he wanted to do if he had remained in school full-time.

Knowing the work he wanted, and having completed his degree emphasizing this field, he was able to pursue a position on a team needing individuals with his specific skills. Having his industry experience and connections, upon completing his degree, he promptly transitioned to a Mission Planning team working on exactly what he wanted to be doing.

Studying Full-Time

If you're the type of personality that doesn't like splitting your focus in too many directions, your best alternative may be to work full-time, then go back to school full-time. Katie Timothy spent two years at a prominent satellite company, enjoyed her time and projects, then decided it was time to head back for her Master's degree.

Having spent two years working, Katie knew exactly what she wanted to study, and what to do with the education when she graduated. For her, taking two years to learn the industry paid off in deciding what to do with her career.

Coming out of her undergraduate degree, she had been accepted to some of the most prestigious engineering programs in the country. When she applied again after working, she received even better offers than her initial round two years earlier.

Charting Your Career

When you go to work your first year, your manager wants to see you begin building a plan for your career. He actually wants to take the time to talk with you about long-term career considerations. He knows that if you have plans and are going places, he can expect you to do good work along the way. He wants you to live up to your ambitions. When it's appropriate, discuss your plans with him. He'll have good advice, and be happy to assist in any way he can.

Ryan Andersen gives us an excellent example of charting your career early on, and involving your managers in the process. The networking path Ryan followed is typical of paths open to many engineers, if they have their eyes open to recognize it. It's worth taking a moment to learn from his experience.

When Ryan was graduating, he interviewed with a young company whose president had graduated from a graduate program which entails both a Master's in Engineering, and a Master's in Business. This was just one of a number of relatively new programs being offered by top universities that have recognized the industry need for business managers with a functional understanding in engineering. Ryan enjoys both engineering and business challenges, so he chose to pursue this program, aware that it requires a few years of experience from applicants.

Ryan started at a large corporation and found his place within his team like every other good engineer. When managers asked about his long-term plans, he would tell them about his ambitions for this graduate program. Ryan did good work, so his managers were fully supportive of his plans. However, he didn't just take a passive interest in getting into the program. Ryan looked further down the road into understanding what he's really interested in: the operational structure of the company.

Within the company, Ryan began actively networking with managers all up and down the chain of command. He didn't start out by saying, "I'm interested in this graduate program." He would inquire about the roles these individuals played, and he would talk about the work he does. It was just normal engineering and business conversation. Naturally, these managers and engineers would eventually ask about his plans for the future. With his goal clearly in mind, it became easy for him to talk about it. From there, it became easy for these managers to support him, and offer to assist in whatever way they could.

In addition to talking with managers in the company, Ryan contacted professors and several alumni of the program. Over two years, he had spoken with nearly every graduate of the program that had some kind of relation to his background and experience. While this networking may not have directly served to get him admitted, he learned much about several industries, and developed a clear vision of what he would gain from the program, and what he wanted to do afterwards.

Ryan doesn't act like a business school poster boy who is bright and full of ambition. He's just another normal guy who went to college and chose to major in Computer Engineering. When you talk with him, he comes across as another everyday, calm, intelligent, level headed engineer.

When Ryan was asked where he learned how to play the networking game, he responded: "Well, I just learned it as I went, I guess. It's not hard. You just talk with people, and then you get better at talking with them."

We've noted here that managers are happy to see a young engineer with ambition, but we need to include a caution with this. Several senior managers have frequently said, "Wearing your ambition on your sleeve is the quickest way to kill your career."

When you express any future ambitions in your plans, you absolutely must be performing in your current responsibilities. Managers will only be happy to see you have ambition if they are confident you are more than capable of achieving your goal. Few things turn people off more than somebody neglecting his current assignment while dreaming of greater responsibility in the future. Ryan was a superb performer in his current job. His managers believed in him, and that's why they supported him and congratulated him when he was accepted to the program he had been aiming toward for so long.

JIT Learning—Continuing Education

There's a concept Dr. Ajit Sabnis, an aerospace technical manager, refers to as *Just in Time learning*, borrowing from the JIT manufacturing principle. This concept applies to all engineers throughout their entire careers.

Dr. Sabnis earned his Ph.D. through pioneering technology in magnetic bearings. He began working with a company on a project closely related to his research. When it was time to lead the team to develop flight hardware, Dr. Sabnis wanted to move from his role as key technologist to that of team manager. Toward that end, he began attending evening classes to acquire the requisite business skills. His career built from there, and he became a very successful team lead and senior technical manager.

Dr. Sabnis continued learning in both the technical and managerial fields. He learned independently, as well as from evening classes. The succession of accumulated classes eventually led to a Master's degree in Executive Management. He recalls that he was actually admitted to the Master's program just one week before receiving his degree.

He has continued on a self-learning path, including simulation and analysis tools in the technical field. All the learning he did was acquired at a time in his career when it was most important to the work he was doing. The needs of the project were what determined his focus of study, so everything he learned immediately went into application. Dr. Sabnis found early in his career that training is most effective if immediately applied, but quickly diminishes if not used. In engineering terms, the half-life of training is extremely small unless extended by rapid, direct use on the job. With training coming *Just in Time* for application, everything Dr. Sabnis learned was effectively and permanently integrated into his overall skill set.

No matter what credentials you start off with, it will always be necessary that you continue learning. As technology advances and markets change, there will never be a time for intellectual stagnation. Continue learning the skills you need for your career, and wherever you go, you will be an effective contributor to every project you work on.

In most scenarios, graduate level education is a smart decision. Whether you go straight through, earn your degree part-time, or return to school full-time, there are advantages and costs to take into consideration. Spend the time to chart your career, and choose the path that is right for you. Regardless of the field or level of your degree, you will continue learning throughout your career. Acquire the necessary skills *Just in Time* for application, and you will be effective in every project you work on.

CHAPTER 13

CONCLUSION

Nuts and Bolts

Engineers are a different breed of individuals. Society has tried to stereotype us over several generations, but once they settle on a stereotype, they find they have to change it. Engineers used to be thought of as mad scientists, social hermits involved in our own technological world. After several years, society realized we're not mad social misfits; we're just interested in things that are really complicated. Then they labeled us as the geeks and the nerds. At that time, geeks and nerds were un-cool. Well, engineers are pretty darn cool. So, now instead of removing the label of geeks and nerds, society just changed the definition to mean smart-cool.

The funny thing is, engineers knew all along that we were smart-cool. It just took the rest of society a few generations to figure it out. If they had asked us at the beginning, they could have avoided all the social confusion about what to label us. But it's no big deal. They seem to have finally come around. Engineers are smart-cool, and we all have a good time.

No single stereotype can be applied to every engineer. Likewise, no single recipe can be followed to lead a successful engineering career. However, we learn from physics that certain principles can always be applied to lead to a predictable outcome. The same holds true with your career. If you apply the right principles in the right situations, you can get off to a good start, and have a career that will be interesting, challenging, and productive.

While building your resume, use your experience to illustrate your interest and understanding. Show the big picture of each project, and follow that with the engineering skills applied to solve design challenges. Real engineering involves a technical project undertaken by a team. Your university is rampant with opportunities for real engineering, so go out there and take it. When your resume is put together, go show it off. Apply through your University Career Services, and apply online.

Look for opportunities to network. You're not looking for a job; you're gaining exposure to know what kind of job you want. Friends from projects and classes in previous years will be happy to tell you about their jobs. Give them a call, and talk about their experiences.

Enjoy every opportunity for an interview. This is your chance to explore the playground real engineers get to work in every day. Take an interest in what the company is doing, and find opportunities to relate your technical experience to what they have in the works. Be ready to teach the interviewer about your projects, illustrating the design factors involved. Talk about the engineering of your whole project, not just the tasks you personally performed. You can illustrate the challenges of your projects with understanding, by identifying the design constraints and the technical solutions you developed. Your portfolio is a very effective tool for communicating about the projects you worked on. This allows interviewers to see your project and the brainwork behind it. Keep your head up, and you'll have a good interview.

When you walk through the door into your first job, be prepared for anything. Your introduction and transition could be smooth, or you could be hanging onto a roller coaster. Whatever happens, remembering the names of the people you meet will make the ride a little more secure. Make a phone call or personal visit for every first contact. Learn to track down people with the information you need.

Apply the art of *Being Interested by Circumstance.* Use your initial assignments to learn the needs of your team. When you get the inevitable stack of documentation, don't *read* the documents, but *learn* the documents and the project. When you have extra time on your hands, invent something useful for your team, instead of enduring the feeling of worthlessness.

Be an Engineer!!! Don't hesitate to understand the systems your team is working on. Jump right into your team, use your engineering mind, and suggest solutions to the challenges they're facing. Remember that one of your primary objectives is learning to do great engineering. Begin developing areas of expertise, and you won't find yourself pigeon-holed.

The most important skill for working in a company is *live communication. Live communication* is the two-way transfer of ideas into and out of the minds of the team members, the ability for getting all points across.

Apply the scientific method of *Guessing What Your Manager Wants.* Ask yourself the right questions, detailed questions that can be answered by yes or no. Look around and see what might be most useful to the team, then inform them that you'd like to start working on that need. Your manager isn't all-knowing, but communicating with him will help him to be adequately-knowing. Suggest solutions, and design solutions, but let your manager and your team know before you *implement* your solutions.

Meetings, documentation, and other paperwork are all necessary tools for engineering. Use these *Office Engineering* tools for their intended purpose, and they will aid your overall company success.

Start engineering your personal finances. Doing so will allow you to truly live comfortably, freeing your mind to be a better engineer.

When considering continuing education, a graduate degree is probably a smart choice. If you already know what you're interested in, a straight path through can be effective.

Working couple years allows you to expand your horizon of interest and earn your degree part-time, or return full-time. Throughout your career, you will always be learning. Seek out *Just In Time Learning* opportunities and you will stay at the top of your game.

When you get down to it, engineers are just a bunch of nuts and bolts. On our own, we may be sort of funny looking, but when you put us all together, we make things work.